Cross-Cultural Design for IT Products and Services

Pei-Luen Patrick Rau • Tom Plocher • Yee-Yin Choong

CRC Press
Taylor & Francis Group
Boca Raton London New York

CRC Press is an imprint of the
Taylor & Francis Group, an **informa** business

CRC Press
Taylor & Francis Group
6000 Broken Sound Parkway NW, Suite 300
Boca Raton, FL 33487-2742

Printed in the United States of America on acid-free paper
Version Date: 20121026

International Standard Book Number: 978-1-4398-3873-0 (Hardback)

Library of Congress Cataloging-in-Publication Data

Rau, Pei-Luen Patrick.
  Cross-cultural design for IT products and services / Pei-Luen Patrick Rau, Tom Plocher, Yee-Yin Choong.
     p. cm. -- (Human factors and ergonomics series)
  Includes bibliographical references and index.
  ISBN 978-1-4398-3873-0 (hardback)
  1. Human-computer interaction. 2. User interfaces--Cross cultural studies. 3. Intercultural communication. 4. Communication in design. I. Plocher, Thomas A. II. Choong, Yee-Yin. III. Title.

QA76.9.H85R38 2012
004.01'9--dc23
                                                              2012009681

Visit the Taylor & Francis Web site at
http://www.taylorandfrancis.com

and the CRC Press Web site at
http://www.crcpress.com

Contents

SECTION III   Methodology

Foreword

This book is published at a very opportune time when there are major activities around the globe including, but not limited to, products and services that are sold worldwide across different cultures, languages, and countries

Industries establish facilities in other countries where they employ local personnel and personnel from the country of origin of the corporation. There are cases of business and diplomatic negotiations in cross-culture settings. For these situations, this book with its 14 chapters, 26 tables, 95 figures, and 276 references provides both the depth and breadth of knowledge to maximize the effectiveness of communication and affective utilization of products and services across different geographic locations. For this purpose, the book effectively illustrates the role of different languages and different cultures within a country and illustrates the effectiveness of cross-culture design by case studies. The chapters conclude with specific guidelines on design of cross-cultural products and services. As such, the book should be of value and great utility to all corporations operating in different languages, cultures, and countries and for researchers working in the cross-cultural area.

Gavriel Salvendy
October 2011

The Authors

Pei-Luen Patrick Rau (BA, Mechanical Engineering, National Taiwan Univer-sity, 1992; MS, Manufacturing, School of Industrial Engineering, Purdue University, 1996; PhD, Human Factors, School of Industrial Engineering, Purdue University, 1998) is a professor in the Department of Industrial Engineering at Tsinghua University in Beijing. He founded and since 2002 has directed the Human–Computer Interaction (HCI) and Usability research center at Tsinghua University, and also directs the Institute of Human Factors and Ergonomics at Tsinghua University. The Institute of Human Factors and Ergonomics at Tsinghua University received the President's Medal from the Ergonomic Society in the United Kingdom (2008). He was named National Science Fund China's Distinguished Young Scholar (2011). His human factors courses were awarded "Excellent Course" from Beijing (2009) and from Tsinghua University. He also received a number of awards from Tsinghua University, including Award for New Academics, Excellent Teaching Achievement, Outstanding Young Teacher award, Excellent Teaching for International Master Program, and Excellent Instruction for Student Research Training. His research themes include cross-cultural design and human-centered design, and he published book chapters and articles on these topics in HCI and related journals and conferences.

Thomas Plocher (MS, Psychology, Yale University, 1975; BA, Psychology, Hamline University, 1973) recently retired from Honeywell International, Inc., and currently works as an industry consultant and educator on human factors and global engineering teamwork. During his 35 years working as a senior human factors engineer at Honeywell, Tom was involved in both research and development. On the development side, Tom applied user-centered design methods to new product development across diverse businesses, including industrial process control, fire systems, video surveillance and security, and health monitoring. In 2005 he was recognized with a Honeywell Technical Achievement Award for his conceptualization and design of Honeywell's FirstVision™ product. He was the first human factors engineer to receive that prestigious award from Honeywell. On the research side, Tom developed numerous innovative technology solutions to difficult human–computer interaction problems, particularly in the areas of hands-free interaction, small screen design, and graphical visualizations of built spaces. His 15 awarded U.S. patents and numerous patents pending resulted from this research. Tom worked with and mentored global engineering and research teams in China since 1998, fueling his interest in cross-cultural collaboration and cross-cultural design. For over 10 years, he has had a highly productive research collaboration with the Institute of Psychology, Chinese Academy of Sciences. He also served as industry advisor and participant in the Cultural Usability (CULTUSAB) research consortium, sponsored by the Danish Research Council. From 2006 to 2009, this multinational team of researchers from Denmark, India, China, and the United States investigated how culture affects usability evaluation and recommended best practices for cross-cultural usability testing. He is the coauthor, with Patrick Rau and Yee-Yin Choong, of several handbook chapters on cross-cultural design of user interfaces.

Yee-Yin Choong (BA, Electronics Engineering, National Chiao-Tung University, Taiwan, 1987; MS, Electronics Engineering, National Chiao-Tung University, Taiwan, 1989; M.Eng, School of Industrial Engineering, Pennsylvania State University, 1991; PhD, Human Factors, School of Industrial Engineering, Purdue University, 1996) is a research scientist at the National Institute of Standards and Technology (NIST) since 2006. Prior to joining NIST, she was a senior Usability Engineer at General Electric for 10 years. During her graduate work at Purdue University, she pioneered the research linking cross-cultural psychology with human–computer interaction (HCI) by studying the impacts of information organization and presentation styles on computer performance of Chinese and Americans due to cultural differences. From her groundwork in cross-cultural HCI research, other researchers have published work that was inspired by her research. Her research has been focused on applying human factors and usability disciplines on any systems with a human-interaction component. The research expands from graphical user interface design, symbol and icons design, to biometrics technology, and cyber security. She has authored and coauthored numerous publications throughout her career.

Section I

Cross-Cultural Psychology

1 A Global Market

1.1 INTRODUCTION

We live in a global marketplace for information technology products and services. The Internet, together with mobile access to information and services, has transformed access to information worldwide. The globalization of companies created an international marketplace enabled by e-commerce. Cloud computing and deploying software as a service allows millions of people to access and use software applications to do their work virtually. Users cross international boundaries to access an enormous range of information.

The global marketplace not only affects how products and services are traded or accessed, but also how they are developed. It has become the norm for an IT product or service to be designed in one country, developed or engineered in a second country, and then traded globally. As almost never before, we are led to question whether the product that emerges from engineering development in that second country accurately reflects the intent of the team that designed it in the first country.

All of the above suggests that, in this new international marketplace, the cultural characteristics of users have become increasingly important. There is an expectation that the design of products and services marketed globally will be sufficiently compatible with local cultures so as to ensure a highly usable and satisfying user experience. The global distribution of design and engineering teams and the markets they serve should cause us to reconsider how we go about validating that design and its engineered realization as a product.

1.2 CULTURAL DIFFERENCES

From cross-cultural psychology, we know that culture has a profound effect on how people think and behave. These differences go far beyond speaking and writing in different languages. We can expect to find cultural differences in values and attitudes, emotional reactions, social relationships, communication styles, visual preferences, and ways of thinking. Physical anthropometry also varies across different world populations, in some cases, quite significantly. The global usability of an information technology product or service will depend, in a large part, on how effectively the design accommodates these psychological and physical differences.

We also are just beginning to understand that usability may not mean the same thing to users in different parts of the world. Granted, there are certain common aspects of usability that everyone values, ease of use being one. However, when we look in a more nuanced way at the characteristics people associate with usability, we find that users in one culture may indeed emphasize and place value on different

3

characteristics than those in another. The design philosophy that guides a new product development must reflect these values. Also, different cultures and subgroups within a culture may have different attitudes toward information technology and be more or less willing to adopt it. The strategy for deploying the product and engaging users to adopt it must be part of both the design philosophy and the product deployment plan.

Designing user interfaces and products for different cultures also affects human factors design methodology. Careful user needs research conducted in the target cultures at the very beginning of a development will ensure that the product, service, or application concept and requirements will support the tasks, work, and lifestyle of the intended users and be compatible with their local environment. The methodology for user needs research must be adapted to the customs, attitudes, and behaviors of the culture being studied. Recent research on cross-cultural usability testing revealed a host of issues inherent in evaluating user interfaces across cultures. Discovering the most important usability problems during an evaluation requires an understanding of these issues and willingness to adapt the test methodology to local cultures.

1.3   ORGANIZATION OF THIS BOOK

Our goal was to write a book on cross-cultural design that would be useful both to researchers and to user-centered design practitioners. For researchers, Section I of this book provides an overview of the dimensions of culture that have implications for human information processing and affective response, and hence, user interface design. For practitioners, Section II provides a set of user interface design guidelines grouped into five areas: language, use of color, icons and images, navigation, and information architecture. Because the usability of many IT products, such as mobile devices, has a significant hardware component, Physical Ergonomics and Anthropometry issues and guidelines are also discussed. The validity of each guideline is discussed in terms of its basis in best practices and standards, human–computer interaction research findings, or cross-cultural theory. We believe this classification will be of interest to both the practitioner and the researcher. For the practitioner, it provides some indication of how firmly the guideline can be argued. For the researcher, this classification will serve to identify topics in cross-cultural design that need further research to be more widely useful.

Section III translates the theory and guidelines into some practical methodology. We discuss the activities and methods of cross-cultural design and, most importantly, how to integrate them into a standard engineering process for product development. We also provide a heuristic evaluation framework that incorporates the cross-cultural design guidelines as criteria and can be used to determine a cross-cultural usability figure of merit for a particular product or user interface design. Finally, we provide suggestions for conducting international user needs research and usability testing.

2 Cross-Cultural Psychology

2.1 FRAMEWORK

The metamodels of culture (del Galdo and Nielsen, 1996; Hoft, 1996; Stewart and Bennett, 1991; Trompenaars, 1993) almost universally conclude that a significant portion of what can be called "culture" is embodied in the psychology of people. This includes their values and attitudes, preferred communication style, and cognitive style. A look at research into these core dimensions of culture provides some insights into how these dimensions might affect user interactions with technology and hence, the design of user interfaces. The role played by values and attitudes and preferred communication style in the framework of culture are discussed below. Culture and cognitive style are discussed in Section 2.2.

2.1.1 VALUES AND ATTITUDES

Hofstede (1991) believed that patterns of thinking, feeling, and acting are mental programs, or, as he dubbed them, software of the mind. These mental programs vary as much as the social environments in which they were acquired. The collective programming of the mind is what distinguishes the values and attitudes of one cultural group from another.

In his classic study of cultural variations in workplace values and attitudes, *Culture and Organizations: Software of the Mind* (Hofstede, 1991), Hofstede reports and interprets his findings from administering his Value Survey Module (VSM) to over 116,000 people in 50 countries, mostly white-collar workers at IBM. Hofstede statistically extracted four dimensions along which his subjects in different national cultures systematically differed: (1) Power Distance, (2) Uncertainty Avoidance, (3) Individualism-Collectivism, and (4) Masculinity-Femininity. Later, he added a fifth dimension: Long-Term Orientation. These are described briefly below.

1. **Power Distance**: Power distance is defined as the extent to which the less powerful members of institutions and organizations within a society accept that power is distributed unequally. Cultures in which high power distance is the norm tend to have highly demarcated levels of hierarchy. People from the lower rungs of the hierarchy have considerable difficulty in crossing the "boundaries." Malaysia, Philippines, India, and Arab countries were examples of high power distance cultures. Austria, Israel, Ireland, New Zealand, and the Scandinavian countries all ranked very low on Hofstede's power distance scale. The United States ranked in the neutral range on this dimension.

2. **Uncertainty Avoidance**: Uncertainty avoidance is the extent to which members of organizations in a society are threatened by uncertainty, ambiguity, and unstructured situations. High uncertainty avoidance cultures tend to have a greater need for formal rules and have less tolerance for people or groups with deviant ideas or behaviors. Latin American countries, Greece, Portugal, Turkey, and Belgium in Europe, and Japan and South Korea in Asia scored high on uncertainty avoidance in Hofstede's research. The Scandinavian countries, Ireland and Great Britain in Europe, Singapore, Hong Kong, and Malaysia in Asia showed the lowest uncertainty avoidance. The United States scored in the middle of the countries sampled.

3. **Individualism-Collectivism**: Individualism describes a society in which the ties between individuals are loose. Everyone is expected to look after himself or herself. Collectivism exists in societies in which people are integrated into strong cohesive groups, which serve to protect them throughout their life. The group receives loyalty in return. The United States, Great Britain, Australia, and Canada had the highest individualism scores in Hofstede's research. The most collectivist countries were in Latin America. Certain Asian countries such as Pakistan, Indonesia, South Korea, and Taiwan also were strongly collectivist.

4. **Masculinity-Femininity**: Masculinity is found in a society in which social gender roles are very distinct. Men are expected to be assertive, tough, and oriented around material success. Women are supposed to be modest, nurturing, and concerned with the quality of life. In feminine cultures, the gender roles overlap. It is OK for both men and women to show traits of nurturing and concern for quality of life. Japan had the highest score for Masculinity in Hofstede's study, followed by Austria, Switzerland, Italy, and Germany in Europe, and numerous Caribbean countries. The United States ranked on the masculinity side of the middle range. The Scandinavian countries and the Netherlands were the most feminine countries studied by Hofstede.

5. **Long-Term versus Short-Term Orientation**: Long-Term Orientation is found in a society oriented toward future rewards. In such societies, perseverance and thrift are valued. Short-term orientation cultures promote the virtues of the past and present. Those include respect for tradition, preserving "face," and fulfilling social obligations. China, Hong Kong, Taiwan, Japan, and South Korea all are good examples for long-term oriented cultures. In Hofstede's survey, Pakistan, Nigeria, Philippines, Canada, Great Britain, and the United States were among the most short-term–oriented cultures.

Since Hofstede's results were originally published, they have been widely used in international management to explain business-related behaviors in different countries (e.g., Trompenaars, 1993). They have been used to understand human-induced causes of airline crashes (Helmreich and Merritt, 1998; Krishnan, Plocher, and Garg, 1999) and medical errors (Helmreich and Merritt, 1998). Krishnan et al. (1999) proposed a range of cockpit user interface enhancements that would potentially counter certain culturally linked attitudinal tendencies that can lead to errors in the cockpit.

Principles relating Hofstede's cultural dimensions and the affect evoked by the visual composition of website content were proposed by Gould (2001). Singh and Matsuo (2004) proposed an extensive list of relationships between Web design features and cultural variables. Finally, much has been written about the implications of these value and attitude dimensions for user interface design (Marcus, 2001).

There is no question that Hofstede's cultural framework can illuminate certain aspects of user interface design, particularly those of an affective or social nature. But as these relationships are postulated and applied to design, one needs to be sensitive to certain issues in applying the framework to new contexts. First, Hofstede believed that his framework tapped into some of the most fundamental and deeply ingrained aspects of national culture. However, he predicted that several of his cultural dimensions, namely Uncertainty Avoidance and Individualism-Collectivism, might be influenced by social, economic, and political change. He envisioned that populations might slowly change over time along these attitudinal dimensions. Results from an application of Hofstede's VSM in China (Plocher et al., 2001) provide some evidence for this. Their engineering student subjects at Tsinghua University scored about as expected on four of the Hofstede dimensions. However, their score on the scale of Individualism-Collectivism was strongly in the opposite direction (strong individualism) to that predicted by the literature. The authors speculated that this particular result may reflect the effects of the rather dramatic social and political upheaval in China over the past 50 years caused by the Chinese revolution and Communist era and, more recently, Western social and economic influences. Young professionals in China may be eschewing the traditional and Communist-era collectivist Chinese values and attitudes, for a more Western-like individualist orientation.

Hofstede's framework provides valuable insights into cross-cultural behaviors but must be applied with an understanding of other influential factors at work in the organization and profession. In a major research project aimed at understanding national, organizational, and professional influences on the team-related behaviors of commercial airline pilots, Helmreich and Merritt (1998) found the Hofstede framework to be extremely useful. Many of the airline crashes they reviewed had a significant cultural component when interpreted within the framework provided by Hofstede. However, Helmreich and Merritt sometimes observed behavior that appeared to contradict their expectations based on national culture alone. They concluded that national culture is only one influence on pilot behavior. In some situations the value and attitudinal dimensions of national culture give way to overriding influences imposed by the culture of the organization (e.g., the airline) or by the piloting profession.

Nevertheless, as Bosland (1985) pointed out, we must remember that Hofstede's cultural dimensions are characteristics of societies that share the same culture. The variables reflect the dominant values of the majority of people living in a given culture. But they do not necessarily reflect the values of every individual person in that culture. Individuals within one culture will have a greater likelihood of holding certain values and attitudes about work than their counterparts in another culture and thus will have a greater likelihood of reacting to events in certain ways. However, within both cultures there will be significant variation, even to the extent of some overlap. Therefore, Hofstede's five dimensions should be understood as formulations of societal attributes rather than individual characteristics.

Self-construal theory focuses more on the individual level of cultural differences. It attempts to explain the different ways in which people construe themselves, both alone and in relationship to others. In the United States a saying such as "the squeaky wheel gets the grease" and in Japan "the nail that sticks up gets pounded down" symbolize meaningful and important cultural differences between the United States and Japan, and between individualistic and collectivistic cultures (Markus and Kitayama, 1991). These differences reflect different concepts of the person as independent from others versus interdependent with others. Self-construal theory is one of the most influential works in culture and psychology; its implications are far-reaching, and its popularity and usage are widespread (Matsumoto, 1999). Table 2.1 summarizes some of the differences between an interdependent and an independent self.

Independent self-construal is the model of the self based on unique individual characteristics. With independent self-construal, "behavior is organized and made meaningful primarily by reference to one's own internal repertoire of thoughts, feelings, and actions, rather than by reference to the thoughts, feelings, and actions of others" (Markus and Kitayama, 1991).

Independent self-construal is similar to Hofstede's (1983) attribute of individualism that emphasized self-esteem, self-identity, and self-image, with personal goals superseding those of the group, and competitive interactions the norm. Interdependent self-construal is the self as defined by relationships with others, especially close others, such as roommates or family members (Markus and Kitayama, 1991). The basis of this self-construal is that the self is "connected to others" (Cross, Bacon, and Morris, 2000) and that relationships are integral parts of the person's well-being. Markus and Kitayama stated that with interdependent self-construal, "behavior is determined and contingent on, and to a large extent, organized by what the actor perceives to be thoughts, feelings, and actions of others in the relationship." This behavior is representative of collectivist cultures described by Hofstede (1983), which are characterized by a rigid social framework with distinct in-group (close family kin) and out-group members, with in-group members conforming to group norms and working together cooperatively. Self-construal plays a central role in one's mental world. For example, it influences, almost always unconsciously, the personal goals and tasks within a society, people's way of evaluating others, and their attitudes toward themselves and others.

Many cultural differences can be ascribed to self-construal. For example, relative to their North American counterparts, individuals in Japan are more likely to rate domains on which they performed inadequately as important and diagnostic of their ability. They did not feel the motivation to trivialize these qualities to restore their perception of themselves (Heine et al., 1999). This finding can be ascribed to the interdependent self-construal that is more prominent in Japanese culture (e.g., Triandis, 1995).

Some other differences can also be ascribed to self-construal. In Japan and China, individuals are more inclined to praise conformity and derogate unique characteristics—characteristics that could challenge the collective. In these cultures, individuals are also more inclined to predict better fortune for other people than for themselves—a pattern that differs from the inclinations of many Americans. Furthermore, in Chinese and Japanese cultures, individuals are especially averse to being conspicuous in their group.

TABLE 2.1

Summary of Key Differences between an Independent and an Interdependent Construal of Self

Feature Compared	Independent Self	Interdependent Self
Self definition	Separate from social context	Connected with social context
Self structure	Bounded, unitary, stable	Unbounded, flexible, variable
Personal goal	Independent from others, express oneself	Interdependent with others
Personal tasks	Be unique	Belong, fit in
	Express self	Occupy one's proper place
	Realize internal attributes	Engage in appropriate action
	Promote own goals	Promote others' goals
	Be direct; "say what is on your mind"	Be indirect; "read other's mind"
Nightmare	Fail to separate	Fail to connect
Perceived important features of a person	Internal, private (abilities, thoughts, feelings)	External, public (statuses, roles, relationships, behavior in situations)
Self-control	Assert inner attributes; change outer aspects	Adjust oneself; change inner attributes
Attitude toward self assertion	Authentic, birthright	Immature, childish for most people; privilege for selected, talented individuals
Role of others	*Self-evaluation*: others important for social comparison, reflected appraisal	*Self-definition*: relationships with others in specific contexts define the self

In China and Japan, individuals tend to focus more on contextual factors rather than individual traits—consistent with their sensitivity to events and constraints in their social environment. They are less likely than are their North American counterparts to ascribe behaviors to the enduring traits of individuals (Markus and Kitayama, 1991; Morris and Peng, 1994).

To explain those cultural differences, the motivational consequences of self-construal are introduced in the next paragraphs. Markus and Kitayama (1991) stated that people with interdependent self-construal experience more interdependent motives than independent people. These motives include deference, the need to admire and willingly follow a superior, to serve gladly; similance, the need to imitate or emulate others, to agree and believe; affiliation, the need to form friendships and associations; nurturance, the need to nourish, aid, or protect another; succorance, the need to seek aid, projection, or sympathy and to be dependent; avoidance of blame, the need to avoid blame, ostracism, or punishment by inhibiting unconventional impulses and to be well behaved and obey the law; and abasement, the need to comply and accept punishment or self-deprecation.

Other empirical studies (Bond, 1986) also found that Chinese respondents show relatively high levels of need for abasement, socially oriented achievement, change, endurance, intraception, nurturance, and order; moderate levels of autonomy, deference, dominance, and succorance; and low levels of individually oriented achievement, affiliation, aggression, exhibition, heterosexuality, and power. However, even before Bond's results, Hwang (1976) found that with continuing rapid social change in China, there was an increase in levels of exhibition, autonomy, intraception, and heterosexuality and a decrease in levels of deference, order, nurturance, and endurance. Interestingly, it appears that those with interdependent selves do not show a greater need for affiliation, as might at first be thought, but instead they exhibit higher levels of those motives that reflect a concern with adjusting oneself so as to occupy a proper place with respect to others.

An interdependent self-construal was demonstrated to enhance the incidence of cooperative, supportive, altruistic, and compliant behavior (e.g., Van Baaren et al., 2004). Both Markus and Kitayama (1991) and Baumeister and Leary (1995) stressed that interdependent relationships are characterized by mutual concern for the interests and outcomes of the other. Experimental research on social dilemmas demonstrated the powerful effect of group identification on participants' willingness to restrict individual gain to preserve a collective good (Brewer and Kramer, 1986). Identification with in-groups can elicit cooperative behavior even in the absence of interpersonal communication among group members. Within the in-group category, individuals develop a cooperative orientation toward shared problems. For example, in a study by Utz (2004) participants were primed with either the independent or interdependent self-construal. They were then asked to participate in a social dilemma. It was found that participants primed with independence showed lower levels of cooperation than participants primed with interdependence (Utz, 2004).

Independent and interdependent construal can result in different values and attitudes toward society and others. Consequently, these can result in different attitudes toward technology. For example, the reasons to adopt a new technology may be different for independent and interdependent people. People who value personal uniqueness might buy a new game console to express the self as modern or specialized, while an interdependent person might buy it mainly because he or she wants to have some common topic to discuss with classmates who also use it. Therefore, self-construal theory can have implications for technology design in many ways, which will be discussed in detail in Chapter 11.

Recently, a third self-construal was observed, which involves the perception of the self as having a deep interconnection with all forms of life or things (DeCicco and Stroink, 2007). This construal differs markedly from the interdependent self, which is only concerned about harmony in relationships with specific other human beings. Metapersonal self-construal explains people's attitudes and behavior toward inanimate things or other forms of life. It was observed that metapersonal self-construal predicts environmental concern, cooperation, and conservation (Arnocky, Stroink, and DeCicco, 2007).

Metapersonal self-construal can explain cultural differences in attitudes toward things. For example, compared with people from other cultures, Japanese are more inclined to perceive things as having life. Consequently, they perceive themselves as interconnected with those anthropomorphic things. This can result in behaviors or expressions that appear strange to other cultures. For example the Japanese have an expression "do not make my teapot lose face." When we design products and services based on many recent technologies, including the Internet and intelligent technologies like anthropomorphic robots, it is important to consider cultural differences in metapersonal self-construal and people's attitudes toward the things surrounding them.

Finally, it should be mentioned that although self-construal is considered as a very stable characteristic in one's life, it does not mean that a cultural emphasis on interdependent or independent or metapersonal will be unchanging. Especially where society experiences tremendous changes in its economics, population, and social structure, China for instance, self-construal tendencies can shift very quickly between generations. Owing to a host of social influences, the generation of young Chinese born in the 1980s and 1990s showed a significant increase in degree of independent self-construal. These influences include the "one-child" policy, fast exposure to the rest of the world, prompt development of modern economic systems such as insurance and banking, and dramatic changes in social structure such as smaller families and more personal responsibility. Therefore, when we analyze the fast-developing cultures, changes should always be highlighted rather than ignored.

In conclusion, Hofstede's five dimensions and self-construal theory provide comprehensive insights of culture values and people's attitudes in each culture type. Hofstede derived his dimensions by measuring societal characteristics at the phenomenon level, while Markus and Kitayama formed the self-construal theory to explain differences between individuals from different cultures.

2.1.2 Preferred Communication Style—Use of Context

Context refers to the amount of information packed into a specific instance of communication (Hall, 1976, 1990). It is one basis for describing communication styles. "High-context" communication style uses terse messages and is short on background details. It assumes that the receiver of the message is familiar with the subject matter. "Low-context" communication style uses more lengthy or elaborate messages, contains a lot of background information on the subject of the communication, and assumes that the receiver of the message may not necessarily be familiar with the subject. When high- and low-context people attempt to communicate, misunderstanding or frustration often results. The low-context person wants more detail and background information than the high-context person is willing or able to provide. The high-context person is impatient listening to "information that he or she already knows" from a low-context person who is only presenting his or her usual complete and thorough message.

Hall believed that communication style, high or low context, was deeply rooted in culture. Although there will be significant variation in style within any one culture, one style will tend to be dominant. Germans, Dutch, English, and Americans tend to prefer low-context communication. French, Italians, Spanish, Latin Americans, and Japanese prefer high-context communication. In their popular book, *Understanding Cultural Differences* (Hall and Hall, 1990), the Halls described and interpreted the stereotypical business-related behaviors of French, Germans, and Americans in terms of communication style and discuss the kinds of conflicts and misunderstandings that can occur when High Context meets Low Context.

Communication style also affects how people interact with information systems, particularly nonlinear, hypertext systems such as the Web (Rau, 2001). High-context people browse information faster and require fewer links to find information than low-context users. However, high-context users also have a greater tendency to become disoriented and lose their sense of location and direction in hypertext. Low-context users are slower to browse information and link more pages but are less inclined to get lost.

2.2 COGNITION AND HUMAN INFORMATION PROCESSING

People interact with the world around them by sensing and perceiving the stimuli presented to them, making sense of the information perceived, deciding if a response is needed, and then executing the response. Many cognitive theories describe how humans perceive (Card, Moran, and Newell, 1983), store, and retrieve information from short-term and long-term memory (Norman and Rumelhart, 1970; Shiffrin and Atkinson, 1969; Shiffrin and Schneider, 1977; Wickens and Hollands, 2000), manipulate that information to make decisions and solve problems (Newell and Simon, 1972), and carry out responses. The stages in human information processing are represented in a general qualitative model by Wickens and Hollands (2000).

The physiological mechanism of human information processing is universal to all people and can be viewed as culturally independent. The organization and structure of the information at each stage are affected by the experience of each individual. One important factor that governs such experience is culture. As noted in the previous section, not only does culture affect the values and attitudes held by people, culture can affect people's cognition as they interact with the world around them.

2.2.1 COGNITIVE STYLE

Since the 1980s, many researchers studied the fundamental differences in cognitive behavior between people of different national cultures. The classic paper by Liu (1986) was the first attempt at describing a Chinese cognitive style and the experiential factors that shape Chinese cognitive style during development: the family order, the Chinese educational system, and the nature of the Chinese language. Hall (1984, 1989, 1990) and Hall and Hall (1990) wrote extensively about time cognition, how it was expressed in many different behaviors of daily life, and how it varied across national cultures. Later, Nisbett (2003) and his colleagues in Asia (Nisbett et al., 2001) developed a significant body of experimental evidence characterizing fundamental differences in reasoning style between Easterners and Westerners.

Perhaps the most clearly understood and documented differences in ways of thinking are the differences between Americans and Chinese. The cognitive style of Americans is inferential-categorical (functional), which means that they have a tendency to classify stimuli on the basis of functions or inferences made about stimuli that are grouped together accordingly (Chiu, 1972). In contrast, Chinese people have a relational-contextual or thematic cognitive style. They tend to classify stimuli on the basis of interrelationships and thematic relationships (Chiu, 1972). The American way of thinking tends to be analytic, abstract, and imaginative. The Chinese way of thinking tends to be synthetic and concrete.

TABLE 2.2
Western and Eastern Reasoning Styles

Western Analytical-Logical Thought	Eastern Holistic-Dialectical Thought
Focus on objects, attributes, categories	Focus on the field surrounding objects; sensitive to covariation in the field; little relevance seen in categories
Apply rules based on the categories to predict and explain the objects' behavior	Little use of universal rules; behavior of an object is explained by situational forces and factors in the surrounding field
Find category learning easy	Find category learning difficult
Organize things functionally (focus on what an object "does")	To the extent they use categories, they prefer to organize things thematically (e.g., use the context or environment as a basis for identifying similarities)
Use formal logic for reasoning, making categories, and applying and justifying rules	Not much use of formal logic; prefer dialectical reasoning such as synthesis, transcendence, and convergence
Eager to resolve contradictions (logic); when logic and experiential knowledge are in conflict, adhere to formal logical rules	Less eager to resolve contradiction; prefer a logic that accepts contradiction
Interpret individual's behavior as a result of their disposition or personality	Interpret peoples' behavior as the result of situational pressures

Nisbett theorizes that these cognitive differences result from the fundamental philosophical differences—Aristotelean versus Confucian—that permeate almost every aspect of these societies from childrearing to education to social structure. According to Nisbett, Westerners tend to be analytical-logical in reasoning style. Easterners tend to be holistic-dialectical. He describes these stereotypical differences in reasoning style in the ways (Nisbett, 2003) shown in Table 2.2.

It is illustrative to consider what can happen when a Western analytical thinker attempts to solve problems or work globally with an Eastern holistic thinker. Two individuals from different cultures might

- Look at the same information or picture and draw different conclusions.
- Pay different attention to minor evidence and be more or less conclusive about their predictions of actions.
- Look at opposition leaders and predict different actions based on their preference for dispositional versus situational explanations.
- Come to different conclusions about the properties of an object when presented with identical evidence from two other seemingly familiar objects.
- Disagree or miscommunicate over how to group things.

2.2.1.1 Organizing and Searching Information

One basic mental process is related to how people group things into categories. Based on the similarities, people categorize objects perceived to have certain characteristics.

People often decide if something belongs in a certain group by comparing it to the representative member of that category. Some categories are universal across cultures. For example, facial expressions that signal basic emotions—happiness, sadness, anger, fear, surprise, and disgust, are similar across cultures, as are certain meanings of colors across culture. People categorize items of information, objects, and functions according to perceived similarities and differences. If the items have two or more attributes, then there is a basis for variation in how they are grouped together. Some people will emphasize a certain attribute and sort items into categories accordingly. Others will focus on another attribute and sort accordingly. Chiu (1972) found that the Chinese prefer to categorize on the basis of interdependence and relationship, whereas the Americans prefer to analyze the components of stimuli and to infer common features. The difference between the analytical and relational thinking styles is mainly based on how subjectivity is treated. The analytical style separates subjective experience by using an inductive process that leads to an objective reality, whereas the relational style of thinking rests heavily on experience and fails to separate the experiencing person from objective facts, figures, or concepts (Stewart and Bennett, 1991).

Choong (1996) conducted research that showed that different cultures often focus on different attributes of the same items or objects. The experimental results provide insights for cross-cultural design. The results showed that Chinese and American users of an online department store performed better if the contents of the store were organized in a manner that was consistent with their natural way of organizing objects—functional for Americans and thematic for Chinese. In other words, Americans would prefer to see products in a department store organized by function: cleaning supplies, linens, and furniture. Chinese prefer to organize products by themes, in the case of a department store, the different rooms of a house: kitchen, bathroom, and bedroom.

Rau, Choong, and Salvendy (2004) conducted a cross-cultural study to compare the impact of knowledge representation (abstract and concrete) and interface structure (functional and thematic) on the performance of Chinese and Americans. Their study provided additional evidence in line with results from previous studies that the Chinese employ a different thinking style from Americans. It also agreed with the previous study (Choong, 1996) that thematic interface structure was advantageous to Chinese users, and especially for cases in which error rate is an important factor in task performance.

Finally, placing objects into categories may simply be less important for people from Eastern cultures. Nawaz et al. (2007) found in a picture sorting task that Chinese made less use of specific categories and more use of the category "other" than Danish users.

So what does this mean for user interface design? Categories form the basis for the information architectures underlying user interfaces of software systems. The content and organization of menus in graphical user interfaces, links in websites, and file directories in most software applications are based on categories. Databases are built on domain ontologies that rigorously apply categories to organize the concepts and objects in a domain. Different cultures will have different ways of categorizing concepts and objects. Furthermore, the concepts, per se, may differ among cultures. Some concepts used in one culture may not even exist in another culture. What happens when a user from one culture, say, the United States, attempts to use a software application that is based on categories (e.g., an information architecture) developed in another culture, such as India? If the information architecture has not been carefully localized to American culture or carefully internationalized to remove the biases of the Indian culture, this will result in products with poor usability. The user will be frustrated as he or she attempts to locate functions in menus or information organized in an unfamiliar way by an original designer or engineer thinking with a different style.

2.2.1.2   Spatial Cognition

Spatial cognition is the process through which individuals gain knowledge of objects and events linked to space (Gauvain, 1993; Mishra, 1997). Frake (1980) analyzed the use of absolute directions and contingent directions in Southeast Asia and California. Frake showed that culture influences the use of directions. The research of Spencer and Darvizeh (1983) concluded that the Iranian children gave more detailed object information on the route way but less directional information compared with British children.

Ji, Peng, and Nisbett (2000) conducted Rod-and-Frame tests to detect the field dependence between Chinese and Americans. They showed that Chinese participants made more mistakes on the Rod-and-Frame test, reported stronger association between events, and were more responsive to differences in covariation, whereas American participants made few mistakes on the Rod-and-Frame test, indicating that they were less field dependent than the Chinese counterparts.

2.2.1.3   Time Cognition

Though time should be technically and objectively the same for everyone, Hall's (1984) classic ethnographic observations showed that different cultures have different attitudes toward time. He classified two ways in which people understand time: monochronic and polychronic time systems. The different attitudes toward time are reflected in many aspects of peoples' lives, including how they adhere to schedules, approach the tasks of their job, and cope with competing task demands (Bluedorn et al., 1999; Haase, Lee, and Banks, 1979; Kaufman-Scarborough and Lindquist, 1999; Lindquist, Knieling, and Kaufman-Scarborough, 2001).

Monochronic time is dominant in Germany, the United Kingdom, the Netherlands, Finland, the United States, and Australia (Hall, 1984; Hall and Hall, 1990). Cultures with a monochronic time orientation treat time in a linear manner. Time is divided into segments that can be easily scheduled and "spent." Monochronic people prefer to follow clear rules and procedures. They prefer to work on one task at a time and are frustrated when other competing tasks disrupt that focus. Monochronic people have narrower views of the overall situation or activity and may miss significant events related to tasks in queue waiting to be served. Clear procedures are important to monochronic users. They are less inclined to invent procedures in new situations or where standard procedures are not available. Monochronic computer users search for information in hypertext in a deliberate and linear manner, making more links than polychronic users to find the same information (Rau, 2001). Hence, they are slower at searching hyperspaces than polychronic users.

Polychronic time is dominant in Italy, France, Spain, Brazil, and India (Hall, 1984; Hall and Hall, 1990). In contrast to monochronic people, polychronic people perceive time in a less rigid, more flexible way. Adhering to rules, procedures, and schedules is not that important to them.

Zhang et al. (2004) found that polychronic users were more inclined to switch back and forth between tasks and applications than were monochronic users. These researchers classified subjects as polychronic or monochronic using the M/P13 scale of Kaufman-Scarborough and Lindquist (1999). A supervisory control task was set up in an industrial training simulator. The simulation presented two controlled variables whose dynamics were different from each other and changed rather quickly. The operator's task was to make manual control inputs to keep both variables within a normal range. Control error and the number of switches between one process and the other were measured. The researchers found that both control error and number of task switches were highly correlated (0.82 and 0.86) with time orientation of the users. Polychronic users had lower error and a greater number of switches. Sample trials from one polychronic and one monochronic subject are shown in Figure 2.1. The figure shows that the monochronic subject focused on process 1 while process 2 went along way out of normal. She made only one switch during the trial which was to make a big correction in process 2 about halfway through. The polychronic subject brought both processes to near normal range right away and continued to keep both processes there by switching frequently between them to make control inputs.

Zhang et al. (2004) conducted a second experiment with a different supervisory control task that required switching attention between tasks running on separate monitors. They recorded eye movements as an indication of task switching. A sample of the eye movement data from one subject of each time orientation persuasion is shown in Figure 2.2. The monochronic subject (top figure) tended to dwell on one screen for a significant time before shifting to the other. The polychronic subject (bottom) moved between the two screens frequently.

When performing searching tasks, the structure of information is very important. Zhao (2002) studied the effect of information structure on performance of information-acquiring tasks by people of monochronic or polychronic time behaviors (see next subsection). She first classified users as monochronic or polychronic by means of a standard survey instrument. Then, she measured their speed of performance on information search and retrieval tasks using two different information structures: hierarchical and network. The hierarchical information structure was basically a tree. It placed information in categories and subcategories, with the vertical structure going up to six levels deep. Crossing over from one vertical branch of the tree to another was possible only at the topmost level. The network information structure allowed more flexible searching based on relationships between items of information, regardless of their place in a hierarchy of categories. The results showed that the monochronic users performed significantly better using a hierarchical information structure. In contrast, polychronic users were significantly faster using a network information structure.

Time orientation, either monochronic or polychronic, is deeply rooted in a culture. Within any one culture, one style tends to be dominant, but there will be significant variation in time orientation and related behaviors. In other words, it is possible

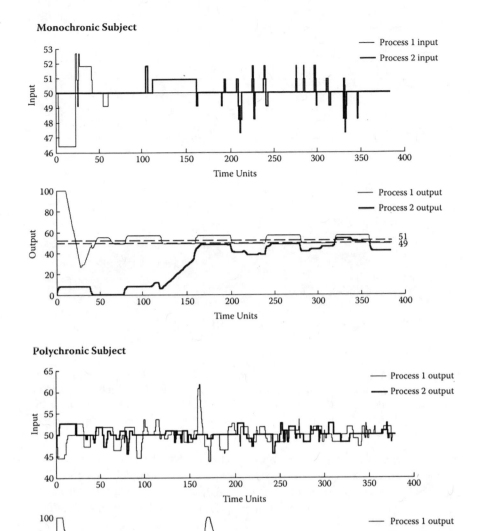

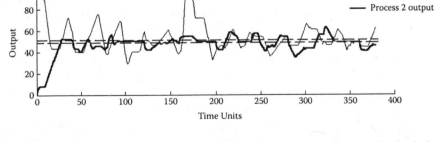

FIGURE 2.1 Sample supervisory control behavior in a process control simulation. Monochronic user shown on top two graphs. Polychronic user shown on bottom two graphs. (From Zhang, Y.,Goonetilleke, R.S., Plocher, T., Liang, S.F.M., 2005. Time-related behaviour in multitasking situations. *International Journal of Human–Computer Studies.* 62(4),425–455.)

Monochronic User

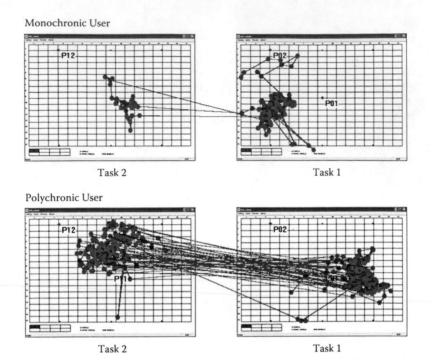

Task 2   Task 1

FIGURE 2.2   Comparison of the eye movements of monochronic (upper graphs) and polychronic (lower graphs) users in a supervisory task that required switching between tasks running on separate displays. (From Luximon, Y. & Goonetilleke, R.S. 2012. Time use behavior in single and time-sharing tasks. *International Journal of Human-Computer Studies.* 70(5),332–345.)

that time cognition can be influenced by factors other than national culture. Zhao, Plocher et al. (2002) found that the natural or preferred time orientation of Chinese industrial workers (monochronic) was quite different from what they displayed on the job (polychronic). During a debriefing of the study, participants revealed that there were many factors at work in the Chinese industrial workplace and in society that simply made polychronic behavior more adaptive.

2.2.1.4   Problem Solving and Decision Making

Problem solving is the process by which people attempt to discover ways of achieving goals that do not seem readily attainable. Cross-cultural research on decision making finds that people of many different cultural groups may use different types of decision-making strategies. Americans may favor considering many possibilities, evaluating each possibility as a hypothesis, and then choosing the best one based on the available information. Other cultures high in uncertainty avoidance, like Chinese, may have a greater tendency to make judgments based on representativeness.

In the research of Yi and Park (2003), more than 800 college students from five countries, Korea, Japan, China, the United States, and Canada, joined the experiments. The researchers found that cultural differences result in different types of decision making. Compared to Americans and Canadians, Korean students showed higher levels of cooperative decision making. Japanese students exhibited the lowest levels of cooperative decision making.

Geary, Fan, and Bow-Thomas (1992) compared the performance of Chinese and American children on the calculation of simple addition. The experiment concluded that Chinese children solved three times more problems correctly than did the Americans. They also did it with greater speed, because Chinese children calculated by the strategies of direct retrieval and decomposition, whereas Americans depended mainly on counting.

2.2.2 LANGUAGE

Every human society, however primitive in other terms, has a language. The ability to use language is perhaps the most profound indicator of the power of human cognition (Miller, 1981). Without language, our ability to remember, to reason, and to solve problems would be severely limited, because so much of human information processing and thinking occurs at the abstract level of language symbols.

Language is key to meaningful communication among people. The communication will be most effective when the first languages or "mother tongues" of the peoples of the world are used. There are between 3,000 and 4,000 spoken languages, with numbers ranging from many millions of speakers down to a few dozen or even fewer. There are hundreds of different written languages represented by scripts in use around the world.

Although a writing system is generally viewed as a system for representing a spoken language, it should be noted that written language is not always a direct transcription of spoken language (Sampson, 1990). For example, in the Arabic-speaking world, vocabulary, grammar, and phonology between spoken and written varieties of Arabic are different. While it is possible to transcribe Arabic speech directly into Arabic script, the transcription will strike Arabic speakers as bizarre and unnatural.

The means to transcribe spoken words are different for different writing systems. For example, Chinese transcribes whole morphemes, the units of meaning. Each single syllabic character represents a unit of meaning that can be a word itself, or sometimes two or more characters form a word. The alphabetic systems such as English transcribe phonemes, the units of sound. There is distinction in accessibility of meanings between morphemic and alphabetical symbols. The meaning of Chinese characters is more manifest perceptually than the meaning of words in alphabetical systems (Hoosain, 1986).

There is evidence showing that language can influence thinking. Hoffman, Lau, and Johnson (1986) asked bilingual English–Chinese speakers to read descriptions of individuals and to provide free interpretations of the individuals described. The descriptions were of characters exemplifying personality schemas with economical labels either in English or Chinese. Bilingual subjects thinking in Chinese used Chinese stereotypes in their free interpretations, whereas those thinking in English used English stereotypes. This indicates that languages can affect people's impressions and memory of other individuals.

Logan (1987) claims that the phonetic alphabet is more than a writing system; it is also a system for organizing information. The alphabet contributed to the development of codified law, monotheism, abstract science, deductive logic, and individualism, each a unique contribution of Western thought. East Asian languages are highly "contextual." Words or phonemes typically have multiple meanings, so they require the context of sentences to be understood. Western languages force a preoccupation with focal objects as opposed to context. For Westerners, the self does the acting; for Easterners, action is something undertaken in concert with others or is the consequence of the self operating in a field of forces (Nisbett, 2003).

REFERENCES

Arnocky, S., Stroink, M., and DeCicco, T. 2007. Self-construal predicts environmental concern, cooperation, and conservation. *Journal of Environmental Psychology*, 27: 255–264.

Baumeister, R.F., and Leary, M.R. 1995. The need to belong: Desire for interpersonal attachments as a fundamental human motivation. *Psychological Bulletin*, 117(3): 497–529.

Bluedorn, A.C., Kalliath, T.J., Strube, M.J., and Martin, G.D. 1999. Polychronicity and the Inventory of Polychronic Values (IPV): The development of an instrument to measure a fundamental dimension of organizational culture. *Journal of Managerial Psychology*, 14:205–230.

Bond, M. H. 1986. *The Psychology of the Chinese People*. New York: Oxford University Press.

Bosland, N. 1985. An Evaluation of Replication Studies Using the Value Survey Model. Tilburg University, Institute for Research on Intercultural Cooperation, the Netherlands.

Brewer, M.B., and Kramer, R.M. 1986. Choice behavior in social dilemmas: Effects of social identity, group size, and decision framing. *Journal of Personality and Social Psychology*, 50(3): 543–549.

Card, S.K., Moran, T.P., and Newell, A.L. 1983. *The Psychology of Human Computer Interaction*. Hillsdale, NJ: Erlbaum.

Chiu, L.H. 1972. A cross-cultural comparison of cognitive styles in Chinese and American children. *International Journal of Psychology*, 7(4): 235–242.

Choong, Y.Y. 1996. Design of Computer Interfaces for the Chinese Population. PhD dissertation, Purdue University, West Lafayette, IN.

Cross, S.E., Bacon, P.L., and Morris, M.L. 2000. The relational-interdependent self-construal and relationships. *Journal of Personality and Social Psychology*, 78: 791–808.

DeCicco, T., and Stroink, M. 2007. A third model of self-construal: The metapersonal self. *International Journal of Transpersonal Studies*, 26: 82–104.

del Galdo, E.M., and Nielsen, J. 1996. *International User Interfaces*. New York: Wiley.

Fernandes, T. 1995. *Global Interface Design: A Guide to Designing International User Interfaces*. Chestnut Hill, MA: AP Professional.

Frake, C. 1980. The ethnographic study of cognitive system. In *Language and Cultural Descriptions*, ed. C. Frake, 1–17. Palo Alto, CA: Stanford University Press.

Gauvain, M. 1993. The development of spatial thinking in everyday activity. *Developmental Review*, 13: 92–121.

Geary, D.C., Fan, L., and Bow-Thomas, C.C. 1992. Numerical cognition: Loci of ability differences comparing children from China and the United States. *Psychological Science*, 3: 180–185.

Gould, E.W. 2001. More than content: Web graphics, cross-cultural requirements, and a visual grammar. *HCI International 2001 Conference*, New Orleans, LA, August.

Haase, R.F., Lee, D.Y., and Banks, D.L. 1979. Cognitive correlates of polychronicity. *Perceptual and Motor Skills*, 49: 271–282.

Hall, E.T. 1976. *Beyond Culture*. New York: Doubleday.

Hall, E.T. 1984. *Dance of Life: The Other Dimension of Time*. Yarmouth, ME: Intercultural Press.

Hall, E.T. 1989. *Beyond Culture*. New York: Anchor Press.

Hall, E.T. 1990. *The Hidden Dimension*. New York: Anchor Press.

Hall, E.T., and Hall, M. 1990. *Understanding Cultural Differences: Germans, French and Americans*. Yarmouth, ME: Intercultural Press.

Heine, S.J., Lehman, D.R., Markus, H.R., and Kitayama S. 1999. Is there a universal need for positive self-regard? *Psychological Review*, 106(4): 766–794.

Helmreich, R.L., and Merritt, A.C. 1998. *Culture at Work in Aviation and Medicine*. Aldershot, UK: Ashgate.

Hoffman, C., Lau, I., and Johnson, D.R. 1986. The linguistic relativity of person cognition: An English-Chinese comparison. *Journal of Personality and Social Psychology*, 51: 1097–1105.

Hofstede, G. 1983. The cultural relativity of organizational practices and theories. *Journal of International Business Studies*, 14(2): 75–89.

Hofstede, G. 1991. *Cultures and Organizations: Software of the Mind. Intercultural Cooperation and Its Importance for Survival*. London: McGraw-Hill International (UK).

Hoft, N. 1996. Developing a cultural model. In *International User Interfaces*, ed. E. del Galdo and J. Nielsen, 41–71. New York: Wiley Computer.

Hoosain, R. 1986. Perceptual processes of the Chinese. In *The Psychology of the Chinese People*, ed. M.H. Bond, 38–72. New York: Oxford University Press.

Hwang, C.H. 1976. Change of psychological needs over thirteen years. *Bulletin of Educational Psychology (Taipei)*, 9: 85–94.

Ji, L., Peng, K., and Nisbett, R.E. 2000. Culture, control, and perception of relationships in the environment. *Journal of Personality and Social Psychology*, 78: 943–955.

Kaufman-Scarborough, C., and Lindquist, J.D. 1999. Time management and polychronicity: Comparisons, contrasts, and insights for the workplace. *Journal of Managerial Psychology*, 14: 288–312.

Krishnan, K., Plocher, T.A., and Garg, C. 1999. Cross-cultural issues in commercial aviation. White Paper. Honeywell Technology Center.

Lindquist, J.D., Knieling, J., and Kaufman-Scarborough, C. 2001. Polychronicity and consumer behavior outcomes among Japanese and U.S. students: A study of response to culture in a U.S. university setting. In *Proceedings of the Tenth Biennial World Marketing Congress*, ed. H. Spotts and H.L. Meadow. Cardiff, Wales, Great Britain, FL: Academy of Marketing Science.

Liu, I.M. 1986. Chinese cognition. In *The Psychology of the Chinese People*, ed. M.H. Bonds, 73–106. New York: Oxford University Press.

Logan, R.K. 1987. *The Alphabet Effect: The Impact of the Phonetic Alphabet on the Development of Western Civilization*. New York: St. Martin's Press.

Luximon, Y. and Goonetilleke, R.S. 2012. Time use behavior in single and time-sharing tasks. *International Journal of Human-Computer Studies*, 70(5) 332–345.

Marcus, A. 2001. Cross-cultural user-interface design. In *Proceeding of the Human-Computer Interface International (HCII) Conference*, ed. M.J. Smith and G. Salvendy, 2: 502–505. Mahwah, NJ: Lawrence Erlbaum Assoc.

Markus, H.R., and Kitayama, S. 1991. Culture and the self: Implications for cognition, emotion, and motivation. *Psychological Review*, 98: 224–253.

Matsumoto, D. 1999. Culture and self: An empirical assessment of Markus and Kitayama's theory of independent and interdependent self-construals. *Asian Journal of Social Psychology*, 2: 289–310.

Miller, G. 1981. *Language and Speech*. San Francisco: Freeman.

Mishra, R.C. 1997. Cognition and cognitive development. In *Handbook of Cross-Cultural Psychology,* Vol. 2, *Basic Processes and Human Development,* ed. J.W. Berry, P.R. Dasen, and T.S. Saraswathi, 143–176. Boston: Allyn and Bacon.

Morris, M.W., and Peng, K. 1994. Culture and cause: American and Chinese attributions for social and physical events. *Journal of Personality and Social Psychology,* 67(6): 949–971.

Nawaz, R., Plocher, T., Clemmensen, T., Qu, W., and Sun, X. 2007. Cultural Differences in the Structure of Categories in Denmark and China, working paper 03-2007, Copenhagen Business School, Copenhagen.

Newell, A., and Simon, H. 1972. *Human Problem Solving.* Englewood Cliffs, NJ: Prentice Hall.

Nisbett, R.E. 2003. *The Geography of Thought: How Asians and Westerners Think Differently and Why.* New York: Free Press.

Nisbett, R.E., Peng, K., Choi, I., and Norenzayan, A. 2001. Culture and systems of thought: Holistic vs. analytic cognition. *Psychological Review,* 108: 291–310.

Norman, D.A., and Rumelhart, D.E. 1970. A system for perception and memory. In *Models of Human Memory,* ed. D.A. Norman, 19–64. New York: Academic.

Plocher, T.A., Zhao, C., Liang, S.M., Sun, X., and Zhang, K. 2001. Understanding the Chinese user: Attitudes toward automation, work, and life. In *Proceedings of the Ninth International Conference on Human-Computer Interaction,* New Orleans, LA, August.

Rau, P.-L.P. 2001. Cross-cultural user interface research and design with emphasis on Asia: Chinese users in Taiwan. *Final Report to Honeywell Singapore Laboratory,* October.

Rau, P.-L.P., Choong, Y.-Y., and Salvendy, G. 2004. A cross culture study of knowledge representation and interface structure in human computer interface. *International Journal of Industrial Ergonomics,* 34(2): 117–129.

Sampson, G. 1990. *Writing Systems: A Linguistic Introduction.* Palo Alto, CA: Stanford University Press.

Shiffrin, R.M., and Atkinson, R.C. 1969. Storage and retrieval processes in long-term memory. *Psychological Review,* 76: 179–193.

Shiffrin, R.M., and Schneider, W. 1977. Controlled and automatic human information processing: II. Perceptual learning, automatic attending, and a general theory. *Psychological Review,* 84: 127–190.

Spencer, C.P., and Darvizeh, Z. 1983. Young children's place descriptions, maps and route finding: A comparison of nursery school children in Iran and Britain. *International Journal of Early Childhood,* 15(1): 26–31.

Stewart, E.C., and Bennett, M.J. 1991. *American Cultural Patterns: A Cross-Cultural Perspective.* Yarmouth, ME: Intercultural Press.

Triandis, H. 1995. *Individualism and Collectivism.* Boulder, CO: Westview.

Trompenaars, F. 1993. *Riding the Waves of Culture.* London: Nicholas Brealy.

Utz, S. 2004. Self-construal and cooperation: Is the interdependent self more cooperative than the independent self? *Self and Identity,* 3(3): 177–190.

Van Baaren, R.B., Holland, R.W., Kawakami, K., and van Knippenberg, A. 2004. Mimicry and prosocial behavior. *Psychological Science,* 15: 71–74.

Wickens, C.D., and Hollands, J.G. 2000. *Engineering Psychology and Human Performance.* Englewood Cliffs, NJ: Prentice Hall.

Yi, J., and Park, S. 2003. Cross-cultural differences in decision-making styles: A study of college students in five countries. *Social Behavior and Personality: An International Journal,* 31(1): 35.

Zhang, Y., Goonetilleke, R.S., Plotcher, T., and Liang, S.F.M. 2005. Time-related behavior in multitasking situations. *International Journal of Human-Computer Studies,* 62(4), 425–455.

Zhao, C. 2002. Effect of Information Structure on Performance of Information Acquiring: A Study Exploring Different Time Behavior: Monochronicity/Polychronicity. PhD dissertation, Institute of Psychology, Chinese Academy of Sciences.

Zhao, C., Plocher, T., Xu, Y., Zhou, R., Liu, X., Liang, S.-F.M., and Zhang, K. 2002. Understanding the polychronicity of Chinese. In *Proceedings of the APCHI2002, Fifth Asia-Pacific Conference on Computer Human Interaction*, 1: 189–196.

Section II

Cross-Cultural Design Guidelines

3 Introduction to Cross-Cultural Design Guidelines

3.1 INTRODUCTION

For some 10 years now we reviewed literature in both Chinese and English languages on international design standards, cross-cultural psychology, and cross-cultural user interface research and design. We reviewed over 200 books, conference papers, and journal articles. Many of these are of a theoretical nature, exploring links between culture, psychology, and human behavior. While fascinating and helpful to us in formulating research questions, these purely theoretical works provide little direct guidance on how to design things for users in different cultures. The majority of the books and papers that we reviewed focused more directly on some specific aspect of cross-cultural user interaction with computers. These research reports usually compared a limited number of cultures, often an Eastern one and a Western one. Thus, the results have a certain amount of validity for the specific cultures studied, but practitioners should be cautious about extrapolating the results too widely. Given the locus of economic development in the world today, it is no surprise that the majority of the research papers investigated users in China, India, and the United States or Northern Europe. That said, most of these papers stopped short of providing clear guidance about design that would be useful in a practical way to a user interface designer. This section of the book attempts to fill that void.

In this section of the book, we attempt to state, in a systematic manner, a set of cross-cultural user interface design guidelines derived from this rather broad literature. We recognize that not all the guidelines presented in this chapter have equal validity. They are derived from a wide range of different sources, from standards and best practices to research findings to theory:

Standards and Best Practices: These guidelines are based on repeated successful design practice, industry best practices, extensive research, or international standards. The books by Fernandes (1995) and del Galdo and Nielson (1996) are good examples.

Research Findings: Numerous experiments and field studies demonstrated cross-cultural differences in performance resulting from particular HCI design features. The work of Shih and Goonetilleke (1998), and Choong and Salvendy (1999) are classic examples.

Theory: Theories such as those of Hall and Hall (1990), Hofstede (1991), and Kress and van Leeuwen (1996) stimulated discussion of cross-cultural design implications. Marcus and Gould (2000), Gould (2001a, 2001b), Gould, Zakaria, and Yusof (2000), and Singh and Matsuo (2004) are examples of practitioners who made thoughtful connections between theory and design practice.

A table is presented in Appendix 1 that shows all the guidelines and how we categorized them according to their basis in practice, research, and theory. Our hope is that this classification will help practitioners apply these guidelines more critically and weight them appropriately in their design practice. But we also hope that our categorization of guidelines will stimulate future research. Any guideline placed in the "theory" category is a topic begging for quantitative research. Kress and van Leeuwen's (1996) "visual grammar" combined with cultural theory is largely unexplored empirically but has huge promise for enhancing the user experience through careful visual-affective design of website content. We just need to validate the principles and practices empirically. The results of such research would most likely move the theory-based guideline into the "research findings" category of validity. And, from there, if the practice of the research-based guideline is successful, it will eventually reach the best practices level.

In the remainder of Section II, guidelines are grouped into chapters on Language (Chapter 4); Color Coding and Affect (Chapter 5); Icons and Images (Chapter 6); Presentation, Navigation, and Layout (Chapter 7); Information Architecture (Chapter 8); and Physical Ergonomics and Anthropometry (Chapter 9). For each guideline, we first present a description of the cross-cultural design problem, citing the relevant literature on cultural differences. Then, the best practice for addressing the design problem, as derived from the literature, is described. Most guidelines are illustrated with an example. Every chapter is concluded with one or more case studies that illustrate a theoretical, methodological, or practical design principle, usually all three.

Note also that in Chapter 15 of Section III on Methodology, a method is presented for using the guidelines to conduct a cross-cultural heuristic evaluation and derive a cross-cultural design figure of merit.

REFERENCES

Choong, Y.Y. 1996. Design of Computer Interfaces for the Chinese Population. PhD dissertation, Purdue University, West Lafayette, IN.

Choong, Y.Y., and Salvendy, G. 1999. Implications for design of computer interfaces for Chinese users in Mainland China. *International Journal of Human-Computer Interaction*, 11: 29–46.

del Galdo, E., and Nielsen, J. 1996. *International User Interfaces*. New York: Wiley.

Fernandes, T. 1995. *Global Interface Design: A Guide to Designing International User Interfaces*. Chestnut Hill, MA: AP Professional.

Gould, E.W. 2001a. More than content: Web graphics, cross-cultural requirements, and a visual grammar. In *Proceedings of the Ninth International Conference on Human-Computer Interaction 2001 (HCI 2001)*. 2: 506–509. New Orleans, LA.

Gould, E.W. 2001b. Using cross-cultural theory to predict user preferences on the web. In *Proceedings of the Ninth International Conference on Human-Computer Interaction 2001 (HCI 2001)*. 2: 546–547. New Orleans, LA.

Gould, E.W., Zakaria, N., and Yusof, S.A.M. 2000. Applying culture to website design: A comparison of Malaysian and U.S. Websites. In *Proceedings of the International Professional Communication Conference (IEEE Professional Communication Society)/ SIGDOC 2000*, 162–171. Cambridge, MA.

Hall, E.T., and Hall, M. 1990. *Understanding Cultural Differences: Germans, French and Americans*. Yarmouth, ME: Intercultural Press.

Hofstede, G. 1991. *Cultures and Organizations: Software of the Mind. Intercultural Cooperation and Its Importance for Survival*. London: McGraw-Hill International (UK).

Kress, G., and van Leeuwen, T. 1996. *Reading Images: The Grammar of Visual Design*. London: Routledge.

Marcus, A., and Gould, E.W. 2000. Crosscurrents: Cultural dimensions and global web-user interface design. *Interactions,* 7: 32–46.

Shih, H., and Goonetilleke, R. 1998. Effectiveness of menu orientation in Chinese. *Human Factors*, 40: 569–576.

Singh, N., and Matsuo, H. 2004. Measuring cultural adaptation on the Web: A content analytic study of U.S. and Japanese Web sites. *Journal of Business Research,* 57(8): 864–872.

4 Language

4.1 INTRODUCTION TO THE PROBLEM OF LANGUAGE

Using text in a user interface imposes significant challenges for cross-cultural design. Just consider the differences between some world languages illustrated in Table 4.1 (modified from Orngreen, Katre, and Datar, 2009).

These differences in language obviously affect the character set required to render the language properly in a user interface. They also affect a host of other things in the user interface, including sizing and spacing of text fields, directionality of menus, font type and size, and text entry medium. The guidelines that follow attempt to capture the range of issues confronting a user interface designer when contemplating the use of text in his or her design. Best practices and software implementation details and advice are offered on-line by IBM (2012b), MicroSoft (2012), Oracle (2010), and W3C (2012).

4.2 DESIGN GUIDELINES

4.2.1 USE SIMPLIFIED ENGLISH

4.2.1.1 Why?

Ambiguous or overly complicated text confuses users, causing frustration, errors, inefficiency, and in some cases, even hazards. For English language screens intended for international use, one solution is to simplify your use of English-language by applying a rigorous editing system such as AECMA (Association Europeenee des Constructeurs de Material Aerospatial) Simplified English. Also, aim for a reading level of U.S. grade level 6.

According to Mills and Caldwell (1997), AECMA Simplified English will

- Reduce the amount of screen text
- Reduce the level of English proficiency needed to read the screens
- Reduce future translation costs
- Reduce the number of reading errors made by nonnative English speakers

4.2.1.2 How?

The rules of Simplified English are

- Use only words and verb tenses approved by the SE (Simplified English) dictionary
- Always use the same word for the same thing

TABLE 4.1

Similarities and Differences between Some World Languages

	English	Danish	Devanagari	Arabic	Chinese
Number of consonants	21	20	36	25	21
Number of vowels	5	9	14	3	16 (spoken Mandarin)
Numerals	10	10	10	10	10
Number of characters	26	29	50	28	3,000–20,000
Cursive style	No	No	Yes	Yes	, No formal cursive system
Character shape	Cubic upper and lower case (Gutenberg style)	Cubic upper- and lowercase (Gutenberg style)	Winding and asymmetrical	Winding and asymmetrical	Ideographic
Ligatures	No	No	Yes, 504 variations of conjuncts	A few	No
Diacritic marks	No	Not compulsory	Yes and necessary	Yes, but few	No
Hyphens and other special characters	Yes	Yes, but seldom	No	Yes	No
Compound words	No	Yes, many	No	No	No
Case	Yes	Yes			No
Directionality	Unidirectional Left to right	Unidirectional Left to right	Unidirectional Left to right	Unidirectional Left to right (Simplified Chinese) Right to left (Traditional Chinese)	
Multiple dialects	No	No	Yes	Yes	Yes

Source: Data from Orngreen, R., Katre, D., and Datar, S. 2009. Displays for textual communication in internationalization and localization perspectives. In *HWID2009*, International Federation for Information Processing AICT 316. Working conference on usability in social, cultural and organizational contexts, in conjunction with the fourth Cultural Usability project seminar, October 7–8, ed. D. Katre et al., 115–131, Pune, India. With permission.

- Use active, not passive, voice verb tenses
- Use short sentences
- Do not remove nouns, articles, and verbs to make sentences shorter (reduces readability)
- When writing procedures, use only one instruction per sentence

The steps involved in applying Simplified English are

- Evaluate each word in the screen text against the vocabulary supplied in the Simplified English dictionary. Know precisely what idea you want to communicate. In Simplified English there is only one word for each idea. Some exceptions may have to be made for vocabulary or jargon associated with a special technical domain (e.g., medical, industrial).
- Eliminate unnecessary words and replace overly complicated words with those recommended by the dictionary.
- Evaluate the reading level of each screen using a software tool such as Correct Grammar™ from Lifetree Software.
- Simplify verb tenses and shorten sentences until the reading level approaches grade level 5 or 6.

4.2.1.3   Example

1. Original paragraph:

Place the water heater in a clean dry location as near as practical to the area of greatest heated water demand. Long uninsulated hot water lines can waste water and energy. Clearance for accessibility to permit inspection and servicing such as removing heating elements or checking controls must be provided.

2. Same paragraph with Simplified English:

Put the water heater in a clean dry location near the area where you use the most hot water. If the hot water lines are long and they do not have insulation, you will use too much energy and water. Make sure you have access to the heating elements and the controls for inspection and servicing.

4.2.1.4   Tools

AECMA (Association Europeenee des Constructeurs de Material Aerospatial) Simplified English. Order from AECMA at: http://www.aecma.org. Use a software tool such as Correct Grammar™ from Lifetree Software to evaluate reading level. IBM (2012a) also offers excellent on-line guidelines about effective writing for translations into other languages.

4.2.2   Use Technology Jargon Words Carefully

4.2.2.1   Why?

In an ancient language such as Chinese, there are few exact equivalents for the high-technology expressions. Also, there will be some variability in translation in different regions, industries, and user groups. Do not assume that professional translators, who are unfamiliar with your product and its functions, will be sensitive to these translation issues (IBM, 2012a; Oracle, 2010).

4.2.2.2 How?

One approach to improve translations of technical jargon is to provide an easy-to-use customization capability with the product. This allows users to create and load their own glossary of idioms to supplement the standard translation of terms provided with the product.

4.2.2.3 Example

A study by Zhang, Zhao, and Zhang (2000) found that more than a quarter of the terms contained in a standard English glossary of petrochemical technology had two or more possible translations into Chinese, depending on the region and the industrial facility. Shown in Table 4.2 is a sample of terms from their glossary for which there were multiple translations into Chinese.

4.2.3   DO NOT USE ABBREVIATIONS

4.2.3.1 Why?

Few abbreviations are truly international in their meaning. Some languages such as Chinese and Devanagari (Hindi) do not have the concept of abbreviations.

4.2.3.2 How?

Avoid using abbreviations in screen text (IBM, 2012a).

TABLE 4.2

Selected Terms from Industrial Glossary Illustrating Difference between the Translator's and the Workers' Interpretation

Terms	Translator's Term	Worker's Idiom
Alarms	警报	报警
Report	报告	报表
Enable/Disable	启用/禁止	允许/禁止
Acknowledge/Silence (alarm)	确认/静声	确认/消音
History Assignment	以往任务	历史任务
Acronyms	缩拼	缩写
Alarm Pager	告警拨叫	报警器
Recipes	处方	配方

Source: Data from Zhang, K., Zhao, C., and Zhang, Tao. 2000. User-Validated Automated Technical Glossary. *CAS Technical Report, Institute of Psychology, Chinese Academy of Sciences*, August 15. With permission.

OFF	OK	ON	STOP	HELP
83%	83%	74%	74%	74%
END	Kg	Mm	DEL	MENU
63%	56%	56%	52%	48%
ESC	EDIT	SHIFT	Rpm	AC
43%	39%	39%	35%	21%

FIGURE 4.1 Recognition of English computer jargon words and abbreviations among Chinese industrial workers. (From Röse, K., Liu, L., and Zühlke, D. 2001. Design issues in mainland China and Western Europe: Similarities and differences in the area of human-machine-interaction design. In *Systems, Social and Internationalization Design Aspects of Human-Computer Interaction*, ed. M.J. Smith and G. Salvendy, 532–536. Boca Raton, FL: CRC Press.

4.2.3.3 Example

Röse et al. (2001) found a surprising lack of recognition for certain English computer jargon words and abbreviations among Chinese industrial machinery operators. The results of her recognition tests are shown in Figure 4.1.

4.2.4 Make Sure Words Are Translated to an Appropriate Context

4.2.4.1 Why?

In some languages, such as Chinese, the same word can be used as a verb and a noun. For example, the translation of "Communications" and "Communicate" is the same. Used as a menu item or label in the user interface, "Communications" refers to a thing or class of things (nouns). "Communicate" implies action initiation (verb). These are very different functions to the user.

4.2.4.2 How?

Make sure your use of language is unambiguous and words are clearly verbs or nouns (IBM, 2012a).

4.2.5 Provide Multiple Language Support

4.2.5.1 Why?

In this global market, many products and applications will be distributed across countries and regions that have different language preferences. Users should have the option of interacting with the product or application in their primary language (IBM, 2012a,b; Microsoft, 2012; W3C, 2012). Luna, Peracchio, and deJuan (2002) point out that websites presented in the user's second language present an additional cognitive processing load. This creates a mismatch between the cognitive effort required to process the site and the amount of cognitive resources or effort the consumer is willing to expend (Peracchio and Meyers-Levy, 1997). They believe that the result, frequently, is a lost consumer of that website.

4.2.5.2 How?

Even users within the same country may prefer different languages or variations in the same language. For example, the languages used in Singapore include English, Mandarin, Malay, and Tamil. American English and British English will have some differences, as well. In portions of Malaysia and Singapore, users sometimes speak a mix of Bhasa and English (Chan and Khalid, 2000).

Therefore, find out if multiple languages are spoken by the target users of your product or application. Even if the target users read and write a single language, determine if there are regional variations. Also, avoid using words from one language as labels on language-selecting controls. For example, it would be bad practice to use the word "Chinese" on a button used to select a Chinese-language option.

4.2.6 MAKE SURE CONTENT MATCHES CONCEPTS AND VALUES OF SELECTED LANGUAGE

4.2.6.1 Why?

De Groot's (1991) theory of language and culture postulates that words map to concepts with culturally specific attributes or features. But the features associated with a word may not be the same in different languages.

4.2.6.2 How?

Luna et al. (2002) proposed a Web design guideline that relates language to website content. They suggest that allowing the user to select a local language is not sufficient. Rather, the content that follows the selection of a local language must be semantically consistent with the concepts and values of the culture. So, a site in which all text was translated into the user's first language may fall short of the mark if the content (images, etc.) linked to the text is not also modified to reflect the user's culture. The user of such a site has to work very hard to connect the text with the culturally less familiar concepts presented in the images of the website. Conversely, a website with text in the user's second language may be rendered more effective if the text is linked to content that reflects the user's own culture, rather than the culture associated with that second language.

4.2.7 ADAPT TO REGIONAL PREFERENCES WHEN DESIGNING SPEECH INTERACTIONS

4.2.7.1 Why?

Interacting with an automated voice is becoming commonplace in today's products. However, there are many facets of the automated voice that are culture sensitive and need to be carefully considered in designing a voice interface.

4.2.7.2 How?

* Consider the gender and tone of the automated voice relative to the gender and culture of the user.

- Consider the need to provide an automated voice that speaks in more than one dialect of the same language.
- The rate of speaking used by the automated voice may need to be slowed down in order to be understood by users in cultures where multiple dialects are spoken.
- Be prepared to design for mixed vocabulary and syntax in speech-driven systems for countries such as Malaysia, Indonesia, and Singapore. Do a careful analysis of vocabulary, syntax, and dialogue requirements, and use an approach such as McCauley's "Multiple Mapping" (McCauley, 1984) to interpret mixed inputs.

4.2.7.3 Example

Many Malaysians mix Behasa Melayu and English. A mixed input to a speech-driven automated teller machine (ATM) system from Chan and Khalid (2000) is shown:

Withdraw	from	savings	account
Withdraw	*dari*	*savings*	*akaun*

4.2.8 Allow Extra Space for Text

4.2.8.1 Why?

In European languages, such as German, words tend to be long strings of characters. They require more space in a box or on a button (IBM, 2012b; Microsoft, 2012; Oracle, 2010; W3C, 2012). In contrast to European languages, Chinese words are short. But the characters are very complex, with up to 20 line strokes in a single character. They require *more pixels or points* than European languages to render clearly on a display screen. Therefore, rendering Chinese characters on a display screen will usually require a larger font size than rendering characters of English or a European language. Also, if you anticipate that the users will have a need to print information from the screen and then transmit it by facsimile, an even larger font size may be called for on the screen to ensure the characters remain clear after printing and faxing.

4.2.8.2 How?

Early in the design process, develop a list of the words and phrases that will be required in the user interface. Translate these into the target languages for the product. Then design the text boxes, menus, and buttons to accommodate the dimensions of the longest strings of characters and the characters requiring the largest font size. If printing/faxing is a requirement, Chinese characters should be large enough to print and fax clearly. Verify it for yourself by trying to print and fax a complex character. Short of this, apply the following rules of thumb:

- European languages:
  - Allow an extra 40% to 50% of line space in text fields, in menus, and on buttons to accommodate future translations into European languages.
  - Avoid narrow columns of text to reduce the amount of word-wrapping in languages characterized by frequent long words.

TABLE 4.3

Differences in Word Length for Some Common Computer Terms in Different Languages

English	File	Edit	View	Print	Help
German	Datei	Editieren	Anzeige	Drucken	Hilfe
French	Fichier	Edition	Visualisation	Impression	Aide
Italian	File	Editare	Visualizzare	Stampare	Aiuto
Spanish	Archivo	Editar	Ver	Imprimir	Ayuda
Chinese (Simplified)	文件	编辑	视图	打印	帮助
Devanagari	संचिका	संपादन	देखें	छापे	सहायता

- Chinese language:
  - Use a minimum font size large enough to occupy a 12 × 12 pixel matrix on the display (Chen, 1989; Tian, 1987).
  - Apply the standard for the People's Republic of China which is a 16 × 14 pixel matrix.

4.2.8.3 Example

Table 4.3 illustrates the differences in word length for a sample of words in several different languages.

4.2.9 Do Not Embed Text in Icons

4.2.9.1 Why?

If text is used in icons, the icon will have to be reprogrammed as part of the translation effort. Recoding icons adds cost and skill requirements to any translation effort. For products that will be translated into other languages, avoid placing text labels on "hard" features such as buttons, keys, and so forth. Such hard labels have to be physically changed (reprinted) as part of the translation process, adding cost and production time (IBM, 2012b; Microsoft, 2012; Oracle, 2010).

4.2.9.2 How?

Design icons that clearly convey their intended meaning without any text. The preferred alternative is to use "soft," programmable buttons and controls in place of hard ones. Then the buttons can be translated simply by changing a text file.

4.2.10 Use an Appropriate Method of Sequence and Order in Lists

4.2.10.1 Why?

For alphabetic languages, such as English, the choices in a list are presented in alphabetical order. However, for languages based on the Latin alphabet, there may be specific collating requirements that are unfamiliar to English-speaking people. Further, some languages, such as Chinese and Japanese, are iconic, not alphabetic. There is no basis in these languages for alphabetical ordering. Therefore, if you must present lists of options in the user interface, then plan to use a method of sequence and order appropriate to the language (IBM, 2012b; Microsoft, 2012; Oracle, 2010).

4.2.10.2   How?

When using languages in the Latin alphabet use the following conventions:

- For Spanish, ñ comes between n and o, and ch needs to be treated as one letter.
- For French, character variants are treated as equivalent, such as (c, ç) and (a, à, á, â).
- Different countries may treat the same character differently. For example, Ä is sorted as equivalent to A in Germany and France; but in Sweden and Finland, Ä is treated as a distinct character and is sorted after Z.

Collating ideographic characters (such as Chinese Hanzi and Japanese Kanji) is more complex than sorting Latin characters. There are four different methods of collating. Select one that is appropriate for your target audience or provide several options for sequencing items in a list and let the user choose one. The four methods of collating in Chinese and Japanese are

- *Radicals*: Radicals are the root forms of a character that give the character its basic meaning. The radical collating sequence sorts according to the radicals that make up the character. If there is more than one character with the same radical, then these similar characters are further sorted by number of strokes that make up the character.
- *Number of Strokes*: Characters are sorted by number of strokes that make up the character. If more than one character has the same number of strokes, these characters are further sorted by radicals.
- *Phonetic Sequence*: Characters are sorted according to the sequence in which they appear in a phonetic alphabet.
- *Frequency of Use*: Place the most frequently used or most important items first followed by decreasingly important items.

For Arabic and Hebrew characters, both single-case languages, there is no collation between uppercase and lowercase characters.

4.2.10.3   Example

Table 4.4 illustrates how collation order of some computer functions varies between English and Mandarin.

4.2.11   AVOID COMBINING USER INTERFACE (UI) OBJECTS INTO PHRASES

4.2.11.1   Why?

It is sometimes convenient to concatenate two or more text-based user interface (UI) objects, such as pull-down lists, so that their entries combine to create meaningful

TABLE 4.4

Difference between English and Mandarin in Sorting Order of the First 10 Windows 7™ Accessories

English	Mandarin (in Taiwan: Sorting According to the First Chinese Character)	Mandarin Translation to English (in Taiwan)	Mandarin (in Mainland China: Sorting in Alphabetical Order)	Mandarin Translation to English (in Mainland China)
Bluetooth File Transfer	Bluetooth 檔案傳輸	Bluetooth File Transfer	Windows 移动中心	Windows Mobility Center
Calculator	Windows 行動中心	Windows Mobility Center	Windows 资源管理器	Windows Explorer
Command Prompt	Windows 檔案總管	Windows Explorer	便笺	Notes
Connect to Network Projector	Wordpad	Wordpad	画图	Paint
Connect to a Projector	小畫家	Paint	计算器	Calculator
Getting Started	小算盤	Calculator	记事本	Notepad
Math Input Panel	同步中心	Sync Center	截图工具	Screenshots tools
Notepad	自黏便箋	Sticky Notes	连接到投影仪	Connect to a Projector
Paint	命令提示字元	Run	连接到网络投影仪	Connect to Network Projector
Run	記事本	Notepad	录音机	Recorder

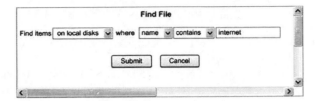

FIGURE 4.2 Example of problematic combination of user interface (UI) objects in a fixed-format sentence.

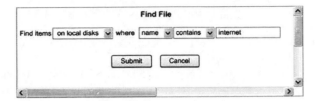

FIGURE 4.3 Preferred design for a "Find File" function into a subject:predicate format.

phrases or sentences. However, such concatenation of text objects is based on the grammatical rules of the particular language being used. In fact, this practice literally embeds those rules for a single language into the user interface. Translated into virtually any other language, the resulting phrases and sentences will not make any grammatical sense.

4.2.11.2 How?

Avoid concatenating text-based UI objects into phrases and sentences (Microsoft, 2012b). Use a "subject:predicate" arrangement for composite phrases or messages.

4.2.11.3 Example

The "Find File" function in Figure 4.2 combines UI objects in a sentence format. Figure 4.3 shows the preferred redesign of the "Find File" function into a "subject:predicate" format.

4.2.12 Avoid the Use of Case as a Distinguishing Feature of Characters

4.2.12.1 Why?

There are a myriad of problems related to including case-sensitive features in your user interface. First, some scripts, such as Chinese, Japanese, Arabic, and Hebrew, simply do not have the concept of upper- and lowercase characters. Second, some programming functions for case conversions will not operate properly on certain scripts. Third, there are problems with converting cases for some scripts, for example:

- Straße (German for "street") becomes STRASSE when capitalized.
- être (French) becomes ETRE when capitalized; however, it is not an issue for Canadian French.

4.2.12.2   How?

Avoid creating case-sensitive features in the user interface (Microsoft, 2012).

4.2.13   TEXT DIRECTIONALITY

4.2.13.1   Why?

Conventional UI design guidelines assume a left-to-right orientation for language. This assumption has extensive implications for screen layout. It recommends left justification for labeling text boxes, presenting text in text boxes, scrolling text within a text box, and presenting a series of control buttons. It also affects menu orientation and the arrangement of items in arrays such as "thumbnails." This convention, of course, assumes that people read left to right. However, numerous written languages are presented in different directions: horizontally or vertically, right-to-left, or even bidirectional.

4.2.13.2   How?

Some examples of languages and their conventions for text directionality (Microsoft, 2012; Oracle, 2010; W3C, 2012) are as follows:

- Horizontal, Unidirectional (Left-to-Right): English, most European languages
- Mixing Horizontal and Vertical Directions: Chinese, Japanese, and Korean support both horizontal and vertical text orientations. For horizontal presentation, it is most common to read/write from left to right, although right-to-left orientation is acceptable. For vertical orientation, it always reads/writes top to bottom, and then flows from right to left. It is common to find a mixture of vertical and horizontal text in newspapers.
- Bidirectional: Languages based on the Arabic and Hebrew scripts are bidirectional. Those languages are read/written from right to left, with embedded Latin text and all numbers running from left to right.
- Vertical Only: Mongolian is an unusual script because it can only be written in a vertical orientation and it reads/writes top to bottom and then flows left to right.

For languages with a right-to-left orientation,

- Place text box label to the right of the box.
- Right justify the text within the box.
- Scroll text in the box so it can be read from right to left.
- Move the insertion point from right to left in front of the leading character.

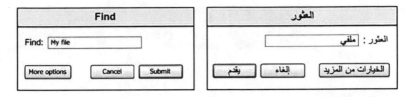

FIGURE 4.4   Text box in English compared to Arabic.

- If two or more buttons are used in the controls, and their order or frequency of use is important, then place them in a right-to-left order, with the most important one placed on the right.
- Use right-to-left orientation to imply the order or sequence of items.
- If an icon is associated with a line of text, position the icon consistent with reading direction (to the right of text item).

4.2.13.3   Example

Figure 4.4 shows a text box in English compared to Arabic.

4.2.14   USE CORRECT LINGUISTIC BOUNDARIES, LIGATURES, TEXT WRAPPINGS, AND JUSTIFICATIONS, PUNCTUATION, DIACRITIC MARKS, AND SYMBOLS

4.2.14.1   Why?

There is tremendous variation between languages in how they separate words and sentences, wrap and justify text, connect characters via ligatures, apply punctuation and diacritic marks, and apply common symbols (Microsoft, 2012; IBM, 2012b).

4.2.14.2   How?

- *Linguistic Boundaries*: Not all languages insert spaces between words; for example, Chinese, Japanese, and Thai do not use any spaces to separate characters. To end a sentence, not all languages use the same symbol (e.g., period "." or question mark "?" as in English).
- *Text Wrappings and Justification*: Text wrappings and justification can follow very different rules for various writing systems. Typical variations in line wrapping include wrapping whole words or breaking the word where it meets the margin (e.g., hyphenate words at the syllabic boundaries). Typical variations in justification include stretching the interword spacing, stretching the intercharacter spacing across the whole line, and stretching the baseline of joined characters. However, the wrapping at whole words or breaking at syllabic boundaries does not always work the same way for all languages using Latin script. For example, in German, hyphenation can actually change the spelling of words, such as "ck" becomes "kk" when split ("Zucker" turns to "Zuk-ker"). For Chinese and Japanese, text can be wrapped anywhere because there are no spaces between words or sentences. However, certain rules need to be followed (called "Kinsoku rules" in Japanese) which specify some characters cannot begin or end a line.

Punctuation			
	Question Marks	Periods	Quotes
Language/ Symbol	French? ¿Spanish? Greek;	English. Japanese 。 Chinese 。 Hindi |	"English" «French» 「 Japanese 」

FIGURE 4.5 A sample of punctuation symbols used by different languages.

Justification is usually unnecessary because the characters are monospaced. Where justification is needed, the intercharacter spacing is stretched.

- *Ligatures*: Languages such as Devanagari make extensive use of ligatures to conjoin consonants into clusters. The designer must consider how the user will interact with the keyboard and system to form these conjunctions.
- *Punctuation*: Be aware that different languages use different punctuation symbols. Figure 4.5 shows several punctuation symbols used by different languages.
- *Diacritic Marks*: In Devanagari, diacritic marks are used to modify the vowel associated with each consonant. The designer must consider how the user will interact with the keyboard and system to indicate these diacritic marks that are so integral to the language.
- *Symbols*: Symbols are characters like the slash (/), the pound sign or number sign (#), the at sign (@), single quotation marks ("), double quotation marks (""), and the ampersand sign (&). In English, these characters are used to mean a variety of things. Whenever you use these symbols, make sure they will be understood by the target users or that they can be translated properly.

4.2.14.3 Examples

Figure 4.5 illustrates a sample of punctuation symbols used by different languages.

4.2.15 CONSIDER LEGIBILITY FACTORS WHEN RENDERING TEXT USING CHINESE CHARACTERS

4.2.15.1 Why?

Chinese characters can be presented in a number of different font styles and with different stroke widths. Legibility varies with the font style and stroke width used. Also, the frequency of the character in the language and the number of strokes in the character affect the accuracy with which Chinese characters can be read on a display screen.

4.2.15.2 How?

Jin, Zhu, and Shen (1988) compared the visual recognition performance of Chinese users for four Chinese font styles: Song, Hei (Black), Zheng Feng Song, and Chang

Fang Song. The results indicated that the successful recognition rates for Song and Hei fonts ranged from about 95% to 98%. Cai, Chi, and You (2001) studied the legibility threshold of Chinese characters in three font styles: Ming, Kai, and Li. The results showed that character style and number of strokes both have a significant impact on the legibility threshold. Ming is the most legible among the three styles, and Kai is significantly more legible than the Li style.

Shen et al. (1988) studied the effect of Chinese character stroke width on reading efficiency. The results indicated that the stroke widths of 0.3 to 0.5 mm for 5 mm height Chinese characters are associated with the highest successful rate of visual recognition. Shieh, Chen, and Chuang (1997) found that characters of higher frequency and fewer strokes were identified more accurately on a visual display terminal (VDT). Also, red-on-green was ranked inferior to color combinations generally used in computer software.

4.2.15.3 Example

More recently, Yang, Xu, and Guo (2011), researchers from China Telecom, conducted research aimed at improving and standardizing the rendering of Chinese characters on small liquid crystal display (LCD) screens. Using data on downloads from leading mobile forums such as Symbian Smartphone Forum, the researchers first identified the most commonly used Chinese fonts. They conducted an experiment to determine which fonts were preferred. The six commonly used fonts are listed in Table 4.5 in the order of user preference.

The researchers conducted further experiments to determine the optimal and smallest acceptable font size for the Song and Yahei fonts displayed on four different LCD screens varying in resolution from 125 PPI to 255 PPI. Their results and recommendations for font size are shown in Table 4.6. Finally, line spacing preferences were studied using a paradigm in which subjects viewed lines of text on a screen and used a slide bar guide to adjust the spacing between the lines. Based on these studies, they recommend that the overall value range for the line spaces on WAP pages should be 131.45% to 181.52%.

4.2.16 SELECT AN EFFICIENT TEXT INPUT METHOD WHEN CHINESE CHARACTERS MUST BE ENTERED

4.2.16.1 Why?

Some languages, such as Chinese, demand special methods for text input. Common input methods use either standardized phonetic notations, such as Pinyin and Zhuyin, or structural information about characters, such as Wubihua or Cangjie. Pinyin requires knowledge of Latinized Chinese Pinyin, which presents difficult barriers to elderly people and nonusers of Pinyin. Wubihua, or Stroke, another popular input method, accepts five distinguishing strokes only when the input sequence is the same as the standard writing order of the character. These methods are difficult to learn, standards vary between different mobile phone manufacturers, and there is significant variation in writing habits between different people.

TABLE 4.5

Samples of Six Popular Chinese Fonts Derived from Frequent Downloads and Shown in Order of User Preference

Font	Sample
Microsoft Yahei	他展主报同现加关外后万面
DynaFont Girl	他展主报同现加关外后万面
Song	他展主報同現加關外后萬面
Yan	他展主报同現加关外后万面
STXingkai	他展主報同現加關外后萬面
Guyin	他展主報同現加關外後萬面

Source: Data from Yang, N., Xu, X., and Guo, F. 2011. A study on the reference standard for WAP-specific Chinese character design. *Presented at the Human-Computer Interaction 14th International Conference, HCI International*, Orlando, FL, July 9–14. With permission.

4.2.16.2   How?

Niu et al. (2010) of Nokia developed a new method called Stroke++ for Chinese character input on mobile phones with the goal of making keypad typing more accessible to novice mobile phone users. The new Stroke++ method exploits the fact that the 600 most frequently used characters in Chinese (out of 20,000 total) can cover 92.9% of the user's needs in short messages, thus greatly reducing the required set of radicals to just 42. These 42 radicals make up most of the keypad. Also, rather than defining rules to restrict the sequence of entry, this method allows users to input radicals in arbitrary order to form a character. The desired character is selected from a pull-down list of candidates generated by the system and organized according to frequency of use. In addition, the keypad layout is designed according to Chinese characters' square shape. The radicals are grouped to make the keypad meaningful. For example, the radicals 金 (Metal), 木 (Wood), 水 (Water), 火 (Fire), and 土 (Earth) representing the five elements in traditional Chinese culture are arranged in a single line to help users to remember. 女 and 人 are put together in the middle to make a Chinese word "woman."

TABLE 4.6

Optimal and Smallest Visible Font Size for Song and Yahei Fonts Displayed on LCD Screens of Various Resolutions

PPI	Font Style	Song	Yahei
125	Optimal	21	20
	Smallest visible	15	15
167	Optimal	24	24
	Smallest visible	18	17
199	Optimal	26	26
	Smallest visible	19	19
255	Optimal	31	32
	Smallest visible	23	23

Source: Data from Yang, N., Xu, X., and Guo, F. 2011. A study on the reference standard for WAP-specific Chinese character design. *Presented at the Human-Computer Interaction 14th International Conference, HCI International,* Orlando, FL, July 9–14. With permission.

4.2.16.3 Example

Figure 4.6 shows the Stroke++ input device for inputting Chinese characters. User performance with the new keypad compared to three other methods is shown in Table 4.7 (Niu et al., 2010). HWR refers to the iPhone's Handwriting Recognition Method.

4.3 CASE STUDY: CROSS-CULTURAL USABILITY ISSUES IN BILINGUAL MOBILE PHONES

4.3.1 BACKGROUND

The inputting and display of text on mobile phone handsets has become of great importance as text-based applications such as SMS/text messaging, e-mail, news, stock reports, and endless other applications have become available to the mobile phone user. However, providing these applications in local languages can be problematic. Cursive-style languages such as Hindi and Arabic, and iconographic languages such as Chinese, are particularly difficult to support in a highly usable manner. The small size and low resolution of the display and limited storage capacity of mobile phone handsets further complicate the problem.

Katre (2006) conducted a heuristic evaluation of mobile phones that were marketed as providing support for both English and Hindi. His evaluation highlights many of the problems confronting the UI designer as he or she incorporates text input and text display capabilities into small format devices. The heuristic evaluation

Input Area

Candidate List

42 Most Common
Radicals

FIGURE 4.6   Stroke++ keypad layout. From Niu, J., Zhu, L., Yan, Q., Liu, Y., Wang, K. 2010. Stroke++: a hybrid Chinese input method for touch screen mobile phones. Paper presented at MobileHCI'10, Lisbon, Portugal.

TABLE 4.7

User Performance with Three Different Text Input Methods Plus One Handwriting Recognition Method

	Success Ratio	Mean Time (Seconds)	Standard Deviation	Percent (%) More Time Than Stroke++
Stroke++	100%	131.92	51.1	—
Pinyin	50%	97.1	29.73	−26%
Wubihua	66%	210.52	127.24	60%
HWR	100%	179.41	97.83	36%

Source: Data from Niu, J., Zhu, L., Yan, Q., Liu, Y., and Wang, K. 2010. Stroke++: A hybrid Chinese input method for touch screen mobile phones. *Paper presented at MobileHCI'10*, Lisbon, Portugal. With permission.

criteria used by Katre are of interest, as well, and are generally applicable to designing text input and displays in other languages.

4.3.2 OBJECTIVES

The objective of Katre's (2006) heuristic evaluation was to improve the state of the art in design and development of more usable mobile phones for Hindi users by

- Providing a set of linguistic usability heuristics
- Using the heuristics to benchmark the text input and text display features in a sample of bilingual English-Hindi mobile phones marketed in India circa 2006

4.3.3 METHOD

The evaluation focused on three bilingual Hindi-English mobile phone handsets common in India around 2006: Nokia/Reliance 3105 CDMA; LG RD5130; and the Samsung/Reliance C200. These are pictured in Figure 4.7.

A set of 10 heuristic criteria was developed and applied in the evaluation of the three handsets. The criteria are

1. Represent the language in its authentic/original form.
2. Maintain the original form and structure of the language's script.

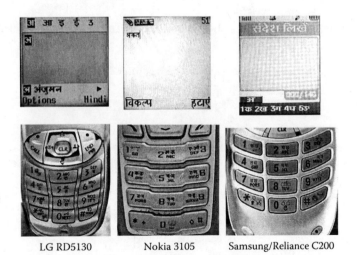

LG RD5130   Nokia 3105   Samsung/Reliance C200

FIGURE 4.7 Three mobile phones evaluated by Katre, 2006. From Katre, D.S. 2006. A position paper on cross-cultural usability issues of bilingual (Hindi and English) mobile phones. In Clemmensen, T. and Yammiyavar, P. (Editors). Indo-Danish Research Symposium 2006-Cross-cultural Issues in HCI, 14-15 May, 2006. Published in India by The Indian Institute of Technology Guwahati, Guwahati 781039, Assam, India. Publication Serial Number: IITG/ DoD / 1 – 5/2006.

3. Represent language uniformly.
4. Avoid influence of English or any other language.
5. Use a maximum of four alphabetic characters per keypad key.
6. Design for the least typing effort possible.
7. Aim for a one-to-one correspondence between keys and characters typed.
8. Avoid uncontrolled mixture and trade-offs between languages.
9. Aim for 100% legibility of text on the display.
10. Strive for high readability/comprehensibility of text.

4.3.4 Differences between English and Devanagari Scripts

Table 4.8 summarizes the basic differences between English and Devanagari scripts.

A look at Table 4.8 quickly suggests why English script is much more easily rendered on the small keyboards and display screens of mobile phones than is Devanagari. English has orders of magnitude fewer alphabetic characters. But even more importantly, the characters are, geometrically, relatively simple. They evolved over the centuries in a very deliberate manner to facilitate block printing and digital rendering. Thus, due to its simplicity, English characters are readily rendered in low-resolution images on small-screen displays and are relatively easy to map to keys on a mobile phone keypad.

In contrast, Devanagari characters are the result of an historically oral communication tradition in India, rather than a print media tradition. For many reasons,

---

TABLE 4.8

Summary of Basic Differences between English and Devanagari Scripts

	Devanagari	English
Number of alphabetic characters	36 consonants, each of which integrates 14 vowels for a total of 504	26
Number of numerals	10	10
Use of upper and lower case	No	Yes
Punctuation and paralinguistic marks	Yes	Yes
Use of acronyms	No	Yes
Basic shapes used	Asymmetric and free-flowing using intricate shapes	Regular grid-based geometric shapes; vertical horizontal, diagonal, and circular
Dynamic change of scripts	Yes, due to conjuncts and matras	No

Source: Data from Orngreen, R., Katre, D., and Datar, S. 2009. Displays for textual communication in internationalization and localization perspectives. In *HWID2009*, International Federation for Information Processing AICT 316. Working conference on usability in social, cultural and organizational contexts, in conjunction with the fourth Cultural Usability project seminar, October 7–8, ed. D. Katre et al., 115–131, Pune, India. With permission.

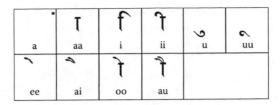

FIGURE 4.8 Set of matras or diacritic marks in Devanagari denoting various vowels. From Katre, D.S. 2006. A position paper on cross-cultural usability issues of bilingual (Hindi and English) mobile phones. In Clemmensen, T. and Yammiyavar, P. (Editors). Indo-Danish Research Symposium 2006-Cross-cultural Issues in HCI, 14-15 May, 2006. Published in India by The Indian Institute of Technology Guwahati, Guwahati 781039, Assam, India. Publication Serial Number: IITG/ DoD / 1 – 5/2006.

they pose major challenges for rendering on low-resolution displays and mapping to mobile phone keypads. Devenngari style of writing is asymmetric, free flowing, and intricate. In addition, its characters are subject to modification by the use of matras and conjuncts that increase the complexity of rendering as text. In Devanagari, each letter represents a consonant, which carries an inherent schwa or *a* vowel. In order to modify this to another vowel, the consonant character is altered by adding a diacritic mark or matra, below, before, or after the consonant to which it refers. Figure 4.8 shows the set of matras denoting the various vowel sounds. As an example, from the character क *ka*, the forms कॆ *ke,* कु *ku,* की *kī,* and का *kā* can be derived, each incorporating a different vowel into the basic consonant character.

Finally, in Devanagari, when consonants occur together without a vowel in between, they are clustered together into conjuncts, which are indicated by special conjunct conventions or ligatures. For example, प + ् + त = प्त. Also, a final consonant is marked with the diacritic ्, called a *halant* in Devanagari. This cancels the inherent *a* vowel, so that, for example, the normal consonant क् नय *knaya* is rendered at the end of a sentence as क् नय् *knay.* As we will see below in the evaluation of mobile phones, the *halant* diacritic mark also is used to denote conjuncts when the use of conjunct ligatures is not feasible.

4.3.5 RESULTS

The results of the evaluation are presented in four parts:

1. Devanagari keypad usability
2. Use of language
3. Character rendering as bitmaps
4. Reading comprehension

4.3.5.1 Devanagari Keypad Usability

Five heuristics focused the evaluation of keypad usability and revealed the problems with mapping characters to keys and typing effort:

Heuristic 2: Maintain the original form and structure of the language's script.

Heuristic 3: Represent language uniformly.

Heuristic 6: Design for the least typing effort possible.

Heuristic 7: Aim for a one-to-one correspondence between keys and characters typed.

Heuristic 9: Aim for 100% legibility of text on the display.

4.3.5.1.1 *Nonstandardized Keypad Layouts and Mappings*

On all English-language mobile phones, numbers and English characters are associated with keys in the same standardized manner. In comparison, the mapping of Devanagari characters was done in a different way on each of the three handsets evaluated here. This lack of standard mapping takes several forms.

First, Devanagari characters are often grouped in small sets to support recall. Each of the three handsets in the study represented these groups differently. Figure 4.9 shows an example and illustrates the usability problem.

There are five characters in this group. Clearly, owing to the small size of the keys, all five cannot be mapped onto one key. So, each brand of phone has chosen a different mapping technique. The LG RD5130 prints the first character of the group on the key and assumes that the user knows the other four. The Nokia handset prints the first and last character in the group and assumes the user knows the three in between. The Samsung phone shows the first two characters of the group. When attempting to type a character, the user must first remember which characters belong to which group and then select that key. Aside from the subset of characters that are actually printed on the keys, there are no hints. If the user does not remember the group in which a character is placed, then he or she has to click through all the keys until he or she finds it. Recall that there are 504 characters in Devanagari, so this can be a daunting search task. Once the correct group is identified, the user presses the key repeatedly until the desired character appears.

A second problem with keypad layouts is that, in some phones, some of the Devanagari characters are completely dropped out to reduce the problem of representing them on keys. These are, ostensibly, those characters that are infrequently used. Further analysis revealed that this criterion was not applied uniformly. In general, out of respect for the linguistic content of a language, this is a poor practice.

FIGURE 4.9 Three different ways of representing a Devanagari character group on a mobile phone key. The group illustrated contains the characters त थ द ध न. From Katre, D.S. 2006. A position paper on cross-cultural usability issues of bilingual (Hindi and English) mobile phones. In Clemmensen, T. and Yammiyavar, P. (Editors). Indo-Danish Research Symposium 2006-Cross-cultural Issues in HCI, 14-15 May, 2006. Published in India by The Indian Institute of Technology Guwahati, Guwahati 781039, Assam, India. Publication Serial Number: IITG/ DoD / 1 – 5/2006.

Another problem highlighted by the evaluation was that Devanagari matras were represented on the keypad in different ways on all three phones. The LG phone mapped them all onto the "9" key. The Nokia phone mapped them to keys 1, 2, or 3. Samsung mapped them to keys 8, 9, or 0. LG and Samsung provided exclusive figures for matras, while Nokia presented them as vowels fused with consonants.

Fourth, two different methods were used for representing conjuncts on the keypad. The LG phone attempted to provide "half characters" for the creation of conjuncts and tried to map them to the keys. The mapping was incomplete and left some half characters unrepresented. Nokia and Samsung used the halant to denote conjuncts which is more true to the Devanagari language conventions.

Fifth, all three mobile phones provided only English-style numeric characters, neglecting the fact that Devanagari has its own set of numeric characters.

4.3.5.1.2 Typing Effort

If too many characters are mapped to each key, the user is forced to make an excessive number of key presses to find the desired character and to type a word or sentence. The Nokia 3105 mobile phone mapped from 6 to 9 Devanagari characters to each key. Katre illustrated this problem by comparing the effort required to type the Devanagari word "Maharashtra" in each of the three phones and an English-language phone. The English-language phone required 20 keystrokes. The worst of the three Devanagari phones was the LG RD5130, which required 55 keystrokes to type this word. One character alone, ष्, required 38 key presses. The resulting word that was produced was also incorrect, read out as "Maharashtta." The LG phone's use of an exclusive set of half characters to form conjuncts forced many additional keystrokes to type conjuncts relative to the other phones. Further, because the half character "r" was missing from the set provided on the phone, the last syllable of the word was displayed incorrectly. Typing "Maharashtra" on the Nokia phone required fewer keypresses, a total of 38, almost entirely due to its use of halant to form the conjunction in the last syllable. This conjunction was typed with 20 fewer keystrokes in the Nokia phone compared to the LG phone. The Samsung phone was the most efficient of all four phones, including even the English-language phone. But it accomplishes this by using a rather arcane set of rules for associating characters with numbers. It is efficient, but not intuitive.

4.3.5.1.3 Attention Span

Mapping an excessive number of characters to each key forces the user to almost continuously press keys. If he or she pauses in between presses, the character associated with that many presses of that key will appear on the screen. But often, with the rapid key pressing, the user skips the desired character. The poor rendering of Devanagari text on the small screen makes it difficult to see what one has selected and further promotes typing errors. Once a typing error is detected, one has to start all over again searching for the desired character.

4.3.5.2 Use of Language on Mobile Phones

On close examination, the user interface of the three Devanagari mobile phones was not always presented in Devanagari script. Heuristic 8—avoid uncontrolled mixture

FIGURE 4.10 Examples of poorly translated technical terms in three mobile phones. From Katre, D.S. 2006. A position paper on cross-cultural usability issues of bilingual (Hindi and English) mobile phones. In Clemmensen, T. and Yammiyavar, P. (Editors). Indo-Danish Research Symposium 2006-Cross-cultural Issues in HCI, 14-15 May, 2006. Published in India by The Indian Institute of Technology Guwahati, Guwahati 781039, Assam, India. Publication Serial Number: IITG/ DoD / 1 – 5/2006.

and trade-offs between languages—highlighted two kinds of problems: language dropouts and use of untranslated English terms.

4.3.5.2.1 Language Dropouts

The evaluation found that some parts of the supposedly Devanagari user interface used English characters. In particular, contact lists and phone numbers were all rendered in English on the three phones.

4.3.5.2.2 Untranslated English Terms

Often, the translation of English technical terms into Devanagari is problematic, resulting in "invented" Devanagari words. Figure 4.10 show screens from the three mobile phones evaluated and examples of poorly translated technical terms on each one.

In the first screen the term "menu" (circled in the lower left corner) is rendered in Devanagari script with no translation. The figure also shows a new term for "caller group," which is replaced on the screen by "Callerkart" which is half English and half Hindi. The same screen also shows "Speed Dial" written in half English and half Hindi. Katre points out that the translation of "speed" as "drutagati" is more appropriate. In total, the evaluation found 42 English terms written in Devanagari script without proper translation on the Nokia 3105 phone. This amounts to about 30% to 40% of the total number of terms in the user interface. These are shown in Table 4.9.

4.3.5.3 Bitmap Fonts and Rendering of Devanagari Characters

The rendering of Devanagari characters is extremely difficult on the low-resolution (128 × 128 pixels) displays used in the phones that were reviewed here. The three phones reviewed here all use bitmap fonts for Devanagari that have several problems:

1. Uneven spacing of characters
2. Distorted shapes of characters
3. Uneven height and width of characters

TABLE 4.9

English Terms Written in Hindi in Nokia 3105

मीनू	Menu	प्रोफाईल	Profile
नंबर	Number	मीटिंग	Meeting
सेटींग	Setting	पेजर	Pager
डायल	Dial	फोन	Phone
टैग	Tag	नेटवर्क	Network
कॉलकर्ता	Caller	कोड	Code
स्क्रॉल	Scroll	पुनडायल	Redial
वी आई पी	VIP	डीटीएमएफ	DTMF
बिजनेस	Business	बैनर	Banner
इनबॉक्स	In Box	गैलरी	Gallery
आइटम	Item	फोल्डर	Folder
इ-मेल	e-mail	ग्राफिक्स	Graphics
स्क्रीन	Screen	मेल बॉक्स	Mailbox
टेम्प्लेट	Template	रिकॉर्डर	Recorder
स्माइलि	Smiley	कॉललॉग	Call Log
छूटी कॉले	Missed Calls	अलार्म घडी	Alarm Watch
कॉलो की सूची	Call List	कैलेंडर	Calendar
ऍनिमेशन	Animation	कैलक्यूलेटर	Calculator
अनलॉक	Unlock	स्टॉपवाच	Stopwatch
कॉलींग कार्ड	Calling Card	कैलोरी गणक	Calorie Counter
डाटा कॉल	Data Call	डाटा दर	Data Rate

Source: Katre, D.S. 2006. A position paper on cross-cultural usability issues of bilingual (Hindi and English) mobile phones. In *Indo-Danish Research Symposium 2006— Cross-cultural Issues in HCI*, ed. T. Clemmensen and P. Yammiyavar, May 14–15. Assam, India: Indian Institute of Technology Guwahati (publication number: IITG/DoD/1–5/2006). With permission.

4. Intersecting or overlapping characters
5. Insufficient spacing between lines of text
6. Ambiguous character shapes
7. Dislocated matras and disjointed conjuncts

These problems violate three heuristics:

Heuristic 2: Represent the language in its authentic/original form.
Heuristic 3: Maintain the original form and structure of the language's script.
Heuristic 9: Aim for 100% legibility of text on the display.

Figure 4.11 shows a sample of Devanagari rendering from each of the three mobile phones, with notable distortion in the shape of the characters in the LG RD 5130 and Nokia 3105 screens shown in Figure 4.12. Note that these distorted characters not only are hard to read, but also they have meanings other than what was intended. Of the three, the Nokia 3105 phone has the most difficulty clearly and accurately rendering Devanagari characters because it attempts to allow the user to view seven lines

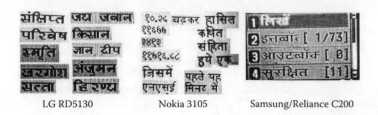

LG RD5130   Nokia 3105   Samsung/Reliance C200

FIGURE 4.11   Rendering of Devanngari fonts in three mobile phones. From Katre, D.S. 2006. A position paper on cross-cultural usability issues of bilingual (Hindi and English) mobile phones. In Clemmensen, T. and Yammiyavar, P. (Editors). Indo-Danish Research Symposium 2006-Cross-cultural Issues in HCI, 14-15 May, 2006. Published in India by The Indian Institute of Technology Guwahati, Guwahati 781039, Assam, India. Publication Serial Number: IITG/ DoD / 1 – 5/2006.

ड़ ॠ   य ॺ   ७ ৪   ५ ६

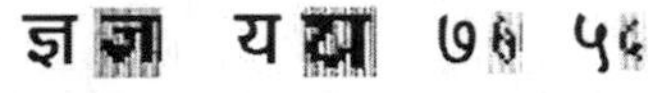

FIGURE 4.12   Distortion of Devanngari characters on the LG RD 5130 (pairs 1 and 2 from the left) and on the Nokis 3105 (pairs 3 and 4 from the left). From Katre, D.S. 2006. A position paper on cross-cultural usability issues of bilingual (Hindi and English) mobile phones. In Clemmensen, T. and Yammiyavar, P. (Editors). Indo-Danish Research Symposium 2006-Cross-cultural Issues in HCI, 14-15 May, 2006. Published in India by The Indian Institute of Technology Guwahati, Guwahati 781039, Assam, India. Publication Serial Number: IITG/ DoD / 1 – 5/2006.

C-DAC सी-डैक C-DAC सी-डैक

FIGURE 4.13   Matching the height of juxtaposed English and Devanngari characters by manipulating font size.

of text at once. However, in order to do this, Nokia has used a font that is too small for accurate and readable Devanagari rendering. The Samsung C200 has the fewest problems with rendering using only four lines of text per screen. This allows it to use a larger font. Auto-scrolling is provided to view more than four lines of text.

Sometimes there is a need to use English and Devanagari characters side by side. Matching the height of juxtaposed English and Devanagari characters can be problematic. Consider the example shown in Figure 4.13. On the left, the English and Devanagari characters both are rendered in 14 point font. The Devanagari appears smaller. On the right, increasing the font of the Devanagari characters to 18 point font makes their height equivalent to the English characters rendered in 14 point font.

4.3.5.4 Reading Comprehension

This area of the evaluation was motivated by Heuristic 10: Strive for high readability/comprehensibility of text. Applying this heuristic to the three mobile phones in the evaluation revealed significant problems with reading and comprehending compound sentences and extensive text passages. In both cases, one has to scroll to read the whole sentence or text passage. The problem lies in the fact that the context for the compound sentence or text passage is usually described in the first few lines of the sentence or passage. Once one scrolls the text, this contextual information disappears from the screen. The user has to remember it in order to comprehend the entire passage. The current evaluation found that users scrolled down to read the entire compound sentence or passage. However, by the time they had read the last sentence, often they had lost the context and had to scroll up to again to view the first lines of the passage. A further problem with scrolled text on mobile phones is that it assumes that users read in a linear manner. Observation of users revealed that their reading of text on a mobile phone is quite nonlinear. Frequently, they scroll back to view earlier sentences or parts of sentences in the passage. Needless to say this is difficult to do on a mobile phone in which barely one simple sentence can be displayed at once on the screen.

4.4 CONCLUSION

The design for use of text on mobile phones requires a significant effort to understand the unique characteristics of the languages to be translated and rendered as part of the user interface. Implications must be considered for keypad design, mapping of characters to keys, text entry, paralinguistic features of the language, correct translation of English technical terms into the local language, mixing of English and local language, rendering and legibility, and text comprehension. Without careful consideration of these issues, usability suffers.

4.5 APPLICATION

Ten heuristic criteria were used here to evaluate three bilingual mobile phones. Applying these criteria revealed a host of important issues about the use of text on mobile phones, all of which affect the usability and acceptability of the product to local cultures. These heuristics were applied here to the use of Devanagari on mobile phones but can be used to guide the design of user interfaces in other languages, as well. Likewise, the types of usability problems identified in this case study should serve as a guide to the user interface designer or content provider faced with localizing the language used in a product.

REFERENCES

Cai, D.C., Chi, C.F., and You, M.L. 2001. The legibility threshold of Chinese characters in three-type styles. *International Journal of Industrial Ergonomics*, 27: 9–17.

Chan, F.Y., and Khalid, H.M. 2000. A method for uncovering the dialog patterns of a speech-driven automatic teller machine for bilingual users. In *Proceedings of APCHI/ASEAN Ergonomics 2000*, 218–223. New York: Elsevier Science.

Chen, J. 1989. Experiments on effects of dot-matrix size and its format of Chinese characters upon VDT visual performance. MA thesis, 356–362, Hangzhou University, China.

De Groot, A. 1991. Bilingual lexical representations: A closer look at conceptual representations. In *Orthography, Phonology, Morphology, and Meaning*, ed. R. Frost and L. Katz, 389–412. Amsterdam: Elsevier.

Goonetilleke, R.S., Lau, W.C., and Shih, H.M. 2002. Visual search strategies and eye movements when searching Chinese character screens. *International Journal of Human-Computer Studies*, 57: 447–468.

IBM. 2012a. Software globalization: writing for an international audience. At: http://www-01.ibm.com/software/globalization/topics/writing/style.html.

IBM. 2012b. Globalize your business: guidelines to design global solutions. At: http://www-01.ibm.com/software/globalization/guidelines/.

Jin, W., Zhu, Z., and Shen, M. 1988. The effect of different typefaces of Chinese characters on recognition. *Engineering Psychology Reports*, 2: 96–100.

Katre, D.S. 2006. A position paper on cross-cultural usability issues of bilingual (Hindi and English) mobile phones. In *Indo-Danish Research Symposium 2006—Cross-cultural Issues in HCI*, ed. T. Clemmensen and P. Yammiyavar, May 14–15. Assam, India: Indian Institute of Technology Guwahati (publication number: IITG/DoD/1–5/2006).

Luna, D., Peracchio, L.A., and deJuan, M.D. 2002. Cross-cultural and cognitive aspects of web site navigation. *Journal of the Academy of Marketing Science,* 30: 397–410.

McCauley, M.C. 1984. Human factors in voice technology. In *Human Factors Review*, ed. F.A. Muckler, 133–166. Santa Monica, CA: Human Factors Society.

Microsoft. 2012. Globalization step-by-step. At: http://msdn.microsoft.com/en-us/goglobal/bb688110.aspx.

Mills, J.A., and Caldwell, B.S. 1997. Simplified English for computer displays. In *Advances in Human Factors and Ergonomics. Design of Computing Systems: Cognitive Considerations. Proceedings of the Seventh International Conference on Human-Computer Interaction*, ed. G. Salvendy, M.J. Smith, and R.J. Koubek. San Francisco, CA, August.

Niu, J., Zhu, L., Yan, Q., Liu, Y., and Wang, K. 2010. Stroke++: A hybrid Chinese input method for touch screen mobile phones. *Paper presented at MobileHCI'10*, Lisbon, Portugal.

Oracle. 2010. I18n in software design, architecture, and implementation. At: http://developers.sun.com/dev/gadc/technicalpublications/articles/archi18n.html.

Orngreen, R., Katre, D., and Datar, S. 2009. Displays for textual communication in internationalization and localization perspectives. In *HWID2009*, International Federation for Information Processing AICT 316. Working conference on usability in social, cultural and organizational contexts, in conjunction with the fourth Cultural Usability project seminar, October 7–8, ed. D. Katre et al., 115–131, Pune, India.

Peracchio, L.A., and Meyers-Levy, J. 1997. Evaluating persuasion-enhancing techniques from a resource matching perspective. *Journal of Consumer Research,* 24: 178–191.

Rau, P.L.P., Plocher, T., and Choong, Y.Y. 2010. Cross-cultural Web design. In *Handbook of Human Factors in Web Design*, 2nd edition, ed. R.W. Proctor and K.P.L. Vu. Boca Raton, FL: CRC Press.

Röse, K., Liu, L., and Zühlke, D. 2001. Design issues in mainland China and Western Europe: Similarities and differences in the area of human-machine-interaction design. In *Systems, Social and Internationalization Design Aspects of Human-Computer Interaction*, ed. M.J. Smith and G. Salvendy, 532–536. Boca Raton, FL: CRC Press.

Shen, M., Zhu, Z., Jin, W., Dai, P., and Han, X. 1988. The effect of Chinese character stroke width on reading efficiency. *Engineering Psychology Research Reports*, 2: 110–114.

Shieh, K.K., Chen, M.T., and Chuang, J.H. 1997. Effects of color combination and typography on identification of characters briefly presented on VDTs. *International Journal of Human-Computer Interaction*, 9: 169–181.

Tian, Q.H. 1987. An experimental study on legibility of the dot-matrix sizes of Chinese characters. BA thesis. Hangzhou University, China.

W3C. 2010. W3C internationalization activity. At: http://www.w3.org/International/questions/qa-mono-multilingual.

Yang, N., Xu, X., and Guo, F. 2011. A study on the reference standard for WAP-specific Chinese character design. *Presented at the Human-Computer Interaction 14th International Conference, HCI International*, Orlando, FL, July 9–14.

Zhang, K., Zhao, C., and Zhang, Tao. 2000. User-Validated Automated Technical Glossary. *CAS Technical Report, Institute of Psychology, Chinese Academy of Sciences*, August 15.

5 Color Coding and Affect

5.1 INTRODUCTION TO THE PROBLEM

There are common physiological bases for color vision (Bond, 1986), so there is no difference between cultures in the actual perception of colors. That said, early studies on color codability demonstrated that people in different societies did not have the same array of colors to partition the color spectrum (Whorf, 1964). Berlin and Kay (1969) argued that if the mechanism underlying color perception is universal, there should be agreement on colors among those who speak different languages from different cultural environments in spite of variations in color vocabulary. They studied some 20 languages and discovered meaningful regularities in the use of basic color terms which are names of color categories consisting of only one morpheme. They also noted an evolutionary progression in color terms in the sense that culturally simpler societies tended to have fewer basic color terms than culturally complex societies, for example large-scale, industrial countries. MacLaury's (1991) work also demonstrated the effect of cultural factors on color coding. A comprehensive study of color naming was presented by Russell, Deregowski, and Kinnear (1997). Davies and Corbett (1997) studied speakers of English, Russian, and Setswana languages and found that they differed in the number of basic color terms and in how the blue-green region is categorized. This suggests that although the strict meanings of basic colors may be similar across cultures, there is still the potential for considerable variability, and it is a good practice to use colors carefully.

5.1.1 COLOR ASSOCIATIONS WITH SAFETY CONDITIONS

5.1.1.1 Why?

Many colors have at least some ambiguity associated with their meaning in different cultures and work contexts. Moreover, operators must learn and remember the meaning of the color codes (Brauer, 1994) which limits the number that can be used to, at most, seven (Wickens and Hollands, 1999). The interpretations of the codes by operators are influenced by their work experiences as well as their cultural background. That is, they tend to decipher the codes based on what they learn. Although there are several sources of reference on the design of display codes for internationalization (e.g., Dreyfuss, 1984; Fernandes, 1995; Miller, Brown, and Cullen, 2000; Peterson and Cullen, 2000), the suggestions are typically more artistic from the perspectives of design and marketing, rather than empirical and data driven. They are not appropriate to guide the use of color in technical or industrial applications to encode safety or alarm conditions.

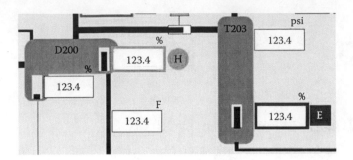

FIGURE 5.1   Example of using shape as a redundant code to color in depicting alarm conditions in an industrial schematic (**see color insert**). Also note the use of gray tones for the plant schematic, with color reserved only for alarm conditions. (Reprinted with permission from Bullemer, Reising, Burns, Hajdukiewicz, and Andrzejewski, Effective Operator Display Design. Abnormal Situation Management Consortium, 2008. http://www.asmconsortium.org/)

5.1.1.2   How?

In technical applications, there is considerable agreement between Asians and Americans about how a very limited set of colors are associated with common safety conditions (Courtney, 1986; Kaiser, 2002; Liang et al., 2000, 2004; Luximon, Lau, and Goonetilleke, 1998). From this research, you can expect that Americans, Chinese, Hong Kong Chinese, Singaporeans, and Malaysians in the industrial setting generally will have the following associations:

- Red—Highest Hazard Level; Orange—High Hazard Level; Green, White, Blue—Lowest Hazard Level
- Red—Highest Temperature; Orange—High Temperature; White or Blue— Lowest Temperature
- Red—Stop; Green—Go

Note that based on existing research, these are the best recommendations we can make for internationalized use of colors to encode safety conditions. But be aware that these associations may not hold in other, non-Western and non-Asian cultures that have not yet been researched and benchmarked for their color associations. Designers should be aware of these ambiguities in color association with safety conditions and always use color coding with caution. The ASM guidelines (Bullemer et al., 2008) provide a good starting point when considering the use of color in operator displays for industrial processes. Also, it is good practice to use a redundant code, such as text or shape, together with color. Figure 5.1 shows an example from the ASM guidelines: red is used to indicate the highest alarm condition and orange a high alarm condition, good international color conventions. But also note that the colors are combined with shapes—a square for the highest alarm and a circle for the high alarm—to provide redundant coding.

5.1.2 Use of Colors That Are Affectively Satisfying in the Target Culture

5.1.2.1 Why?

What people feel about colors is more subject to cultural variation than technical associations with color. Colors and the combination of colors have different affective meanings in different cultures, often quite potent. Many researchers conducted studies on color and its impact for product design.

5.1.2.2 How?

As a general principle, if the same color is to be used across locales, it is a good idea not to use primary colors in design because they may carry negative or positive meanings in some cultures (Osgood, May, and Miran, 1975). Also, there are numerous excellent resources that deal with color usage. Jill Morton's colormatters.com website has extensive information on color studies and online books about color.

5.1.2.3 Example

Minocha, French, and Smith (2002) conducted informal observations and analysis of the choice of colors on some e-Finance sites in India and Taiwan. The website of the ICICI Bank, the second largest commercial bank in India and a pioneer in Internet banking, was selected as most representative of Indian color preferences and the most skilled in using those to communicate a positive affective experience to users. The home page of the ICICI Bank is shown in Figure 5.2. Note the use of red and saffron colors. For Indian users, use of red is associated with vitality, energy, prosperity, and health. Also, red is considered stimulating and shows ambition and initiative. In religious ceremonies and marriages, the guests dress in red clothes. The use of saffron is considered auspicious among Hindus, Sikhs, Jains, and Buddhists. The combination of red and saffron can be considered to signify prosperity and growth for current and prospective customers.

5.2 CASE STUDY

A series of studies conducted by Honeywell's Singapore Laboratory from 2000 to 2003 set out to determine if industrial control room operators from different cultural groups in Asia interpret color codes, shape codes, and layout orientations in the same way (Liang et al., 2000, 2004). Their aim was to identify best practices for the use of display codes and layout directionality when attempting to internationalize process control displays. This case study will focus on their research with color codes for temperature conditions and hazard levels.

5.2.1 Methodology

This field survey was conducted as three separate studies, each in a different country—China, Singapore, and Malaysia. The initial study, conducted in China, used a paper-based questionnaire. Subsequent studies in Singapore and Malaysia used a computerized version of the same rating scales and rating tasks. In all three studies,

FIGURE 5.2   The website of the Indian ICICI bank uses colors (**see color insert**) effectively to give users a positive affective experience. (From French, T., Minocha, S., and Smith, A. 2002. eFinance Localisation: an informal analysis of specific eCulture attractors in selected Indian and Taiwanese sites. In: Coronado, J., Day, D., Hall, B. (Eds.), Designing for Global Markets, Proceedings of IWIPS 2002, vol. 4. Products and Systems International, pp. 9–21.)

the survey was conducted on-site at the operators' workplace by a researcher fluent in both English and the local language. The survey was available in three languages, namely Malay, Chinese, and English. Subjects completed the survey in their preferred language.

5.2.2 DEMOGRAPHICS

5.2.2.1   Malaysia

266 industrial workers participated in the Malaysia study. They came from various industrial sites in Malaysia: petrochemical, gas, and natural oils industries. The age of participants ranged from 19 to 54 years with a mean age of 32.3 years. The subjects were recruited from both East Malaysia ($n = 151$) and West Malaysia ($n = 115$). Subjects were given the language option, and 135 answered the questionnaire in Malay, while 131 answered in English. The ethnic components of the sample reflected the diversity of the Malaysian population. About 4% of participants were Malays, 16% were Chinese, 6% were Indian, and indigenous peoples (e.g., Iban, Bidayuh, and Melanau) and other ethnic groups made up the rest of 33%. About 99% of the subjects' current nationality and nationality at birth was Malaysian. About 81% of the subjects had formal education between 11 and 15 years, 8% had 10 years or less, and 11% had 16 years or more. In terms of work experience in industry, about 22% had 3 years or less, 19% were between 4 and 7 years, 2% were between 8 and 11 years, 8% were between 12 and 15 years, and 26% were 16 years or over.

5.2.2.2 Singapore

The 30 participants in the Singapore study were process control operations students at Temasek Polytechnic School in Singapore. Twenty-three were in their second year of this 2-year program, and seven were in their first year. All the subjects were between 17 and 22 years old with the median at 19 years old.

5.2.2.3 China

Participants in the study came from four industrial plants in the Tianjin region of China, including oil and gas refining, and acrylic and chemical fiber production. The 30 participants included 15 plant engineers, 12 plant operators, and 3 plant managers. This was a rather well-educated group of industry personnel, with 21 of the 30 having completed some college or university studies and all having completed at least high school and technical school. Twenty-three out of the 30 had automated control system experience, mostly with the Honeywell TDC 3000™ product.

5.2.3 COLOR ASSOCIATION QUESTIONS

Nine colors were selected for the studies: red, orange, gray, yellow, green, blue, purple, white, and black. The colors were presented one at a time, either on paper in the China study or on the top center of the computer screen in the other two studies. A scale was provided for rating the temperature level associated with each color and presented under the color to be rated. The color scale had five benchmarks, ranging from Cold, Cool, Warm, to Hot. Not Cool/Not Warm provided the benchmark in the middle of the scale. A scale was also provided for rating hazard level associated with each color. It had five hazard benchmarks ranging from Very Safe, Safe, Neutral, Dangerous, to Very Dangerous. Subjects indicated their choices by circling (in the China study) or pointing and clicking (Singapore and Malaysia) on one of the scale benchmarks. They then clicked the "done" button at the right bottom corner of the screen to proceed to the next question. The presentation order of the nine colors was randomized for each participant. Figure 5.3 shows a sample color association question.

5.2.4 RESULTS

The results from the Wilcoxon Rank Sum tests performed on the color association data are shown in Figure 5.4 and Figure 5.5 for temperature and hazard level associations across the three countries. For each country, the colors are shown from left to right in the order of their mean association preference. The underline indicates mean ratings that are not significantly different at $p < 0.05$.

With respect to the color–temperature association, the results from all three countries were remarkably consistent at the high temperature end of the scale. Red represented the highest degree of temperature, followed by orange. Malaysian and Chinese subjects also associated yellow with a high but lesser temperature than red or orange. At the low temperature end of the scale, white and blue both represented the lowest temperatures in the Singaporean and Malaysian samples. But Chinese preferred to use black to represent the lowest temperature. Color associations are much less clear for middle-range temperatures, and preferences varied somewhat between

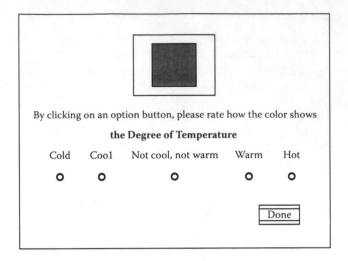

FIGURE 5.3   A sample color question (**see color insert**). (From Liang, S.-F.M., Khalid, M., Taha, Z., and Plocher, T. In search of internationalized operator interface displays in process control: A comparison among Malaysian, Singaporean and Chinese. WWCS, 2004. With permission.)

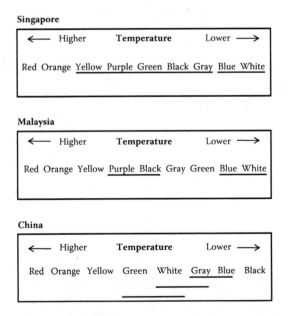

FIGURE 5.4   Survey results comparing color associations with temperature in Singapore, Malaysia, and China. (From Liang, S.-F.M., Khalid, M., Taha, Z., and Plocher, T. In search of internationalized operator interface displays in process control: A comparison among Malaysian, Singaporean and Chinese. WWCS, 2004. With permission.)

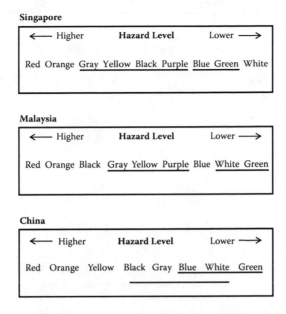

FIGURE 5.5 Survey results comparing color associations with hazard levels in Singapore, Malaysia, and China.

countries. Depending on country, a variety of colors—purple, yellow, gray, green, white, and black—appear to be somewhat equivalent, as indicated by the Wilcoxon tests, in their association with middle-range temperatures.

In terms of color association with hazard level, the results from these three different culture groups showed a significant amount of consistency. Red was the unanimous color to represent the highest hazard level, followed by orange. White was the preferred color in all three country samples to represent the lowest hazard level. However, in the Malaysian sample, white was not significantly more potent than green, and in the Chinese sample, not significantly more potent than either green or blue. Yellow is often associated with "caution" in the West, but here its association with a mid-to-high level of hazard was inconsistent. Only the Chinese participants associated yellow with somewhat less hazard than orange. Malaysian subjects preferred black. The Singaporean subjects chose both yellow and black to represent the same hazard level. Color associations were much less clear for middle hazard levels, and preferences varied somewhat between countries. Depending on country, some combination of gray, purple, black, yellow, and blue was associated with mid-range hazard level. Within these mid-range associations for each country, preferences for one color over another were not significant.

5.3 CONCLUSION

This study of color association with temperature and hazard level among industrial process workers in three Asian countries showed that only a few colors have clear, consistent associations. Red and orange appear to be clearly associated with highest

TABLE 5.1

Summary of Color Associations with Temperature and Hazard Level for Chinese, Malaysian, and Singaporean Samples

	Chinese	Malaysian	Singaporean
	[Red]	[Red]	
Temperature	[Orange]	[Orange]	[Red]
Higher	[Yellow]	[Yellow]	[Orange]
↕	↕	↕	↕
Lower	[Gray and Blue]	[Blue]	[White and Blue]
	[Black]	[White]	
Hazard	[Red]	[Red]	[Red]
Level	[Orange]	[Orange]	[Orange]
Higher	[Yellow]	[Black]	↕
↕	↕	↕	[Blue and Green]
Lower	[Blue, White, and Green]	[Blue, White, and Green]	[White]

and high temperatures, respectively. Blue and white are associated with lowest temperatures, except for Chinese participants, who preferred black. Perhaps in China, black is a metaphor for "total lack of heat," but we can only speculate on that. Red and orange also have clear associations with highest and high hazard levels across all three country samples. Blue, white, and green consistently signaled the lowest or low hazard levels. The rest of the colors tested in these surveys—purple, yellow, black, gray— generally were associated with mid-range temperature and hazard levels and did not differ significantly from one another. The overall comparisons of color associations among the three countries, China, Malaysia, and Singapore, are summarized in Table 5.1.

5.4   APPLICATION

The results of this study should motivate user interface designers to use color carefully when signaling temperature or hazard level. Even within this relatively homogenous sample—all Asian countries and all refining industry workers or trainees—there was a large amount of ambiguity in participants' associations of many of the colors tested. That said, a few colors such as red and orange, and to a slightly lesser extent, blue, white, and green, appear to have consistent and reliable meanings. These could be used as signal colors for temperature and hazard levels in these three countries. Using them in combination with text labels and icons would ensure that even these are not misinterpreted.

REFERENCES

Berlin, B., and Kay, P. 1969. *Basic Color Terms.* Berkeley and Los Angeles: University of California Press.

Bond, M.H. 1986. *The Psychology of the Chinese People.* New York: Oxford University Press.

Brauer, R.L. 1994. Visual environment. In *Safety and Health for Engineers,* Chap. 20, 297–306. New York: Van Nostrand Reinhold.

Bullemer, P., Reising, D., Burns, C., Hajdukiewicz, J., and Andrzejewski, J. 2008. Effective Operator Display Design. Abnormal Situation Management Consortium. www.asmconsortium.org

Courtney, A.J. 1986. Chinese population stereotypes: Color associations. *Human Factors,* 28(1): 97–99.

Davies, I.R., and Corbett, G.G. 1997. A cross-cultural study of colour grouping: Evidence for weak linguistic relativity. *British Journal of Psychology,* 88: 493–517.

Dreyfuss, H. 1984. *Symbol Sourcebook: An Authoritative Guide to International Graphic Symbols.* New York: Van Nostrand Reinhold.

Fernandes, T. 1995. *Global Interface Design: A Guide to Designing International User Interfaces.* Chestnut Hill, MA: AP Professional.

French, T., Minocha, S., and Smith, A. 2002. eFinance localisation: An informal analysis of specific eCulture attractors in selected Indian and Taiwanese sites. In: Coronado, J., Day, D., and Hall, B. (Eds), Designing for Global Markets. Proceedings of IWIPS 2002, vol. 4. Products and Systems International, pp. 9–21. Austin, Texas, USA.

Kaiser, J. 2002. Web Design and Color Symbolism. http://webdesign.about.com/mbody.htm

Liang, S.-F.M., Khalid, M., Taha, Z., and Plocher, T. 2004. In search of internationalized operator interface displays in process control: A comparison among Malaysian, Singaporean and Chinese. WWCS.

Liang, S.-F.M., Plocher, T.A., Lau, P.W.C., Chia, Y.T.B., Rafi, N., and Tan, T.H.R. 2000. Perception of colors and graphics in process control workstations. Proceedings of APCHI/ASEAN Ergonomics 2000, 120–124.

Luximon, A., Lau, W.C., and Goonetilleke, R.S. 1998. Safety signal words and color codes: The perception of implied hazard by Chinese people. Paper presented at the 6th Pan-Pacific Conference on Occupational Ergonomics. Kitakyushu, Japan 1998 (5th PPCOE). Information source: http://www.cimerr.net/aje/v1n1/editorial.html.

MacLaury, R.E. 1991. Social and cognitive motivations of change: Measuring variability in color semantics. *Language,* 67(1): 34–62.

Miller, A.R., Brown, J.M., and Cullen, C.D. (2000). *Global Graphics: Symbols: Designing with Symbols for an International Market.* Minneapolis, MN: Rockport.

Morton, J. 2003. Color Matters. http://www.colormatters.com. (Accessed October 1, 2010.)

Osgood, C.E., May, W.H., and Miron, M.S. 1975. *Cross-Cultural Universals of Affective Meaning.* Urbana, IL: University of Illinois Press.

Peterson, L.K., and Cullen, C.D. 2000. *Global Graphics: Color: Designing with Color for an International Market.* Minneapolis, MN: Rockport.

Russell, P.A., Deregowski, J.B., and Kinnear, P.R. 1997. Perception and aesthetics. In J.W. Spartan. 1999. Multimedia presentation on the colors of culture. http://www.mastep. sjsu.edu/Alquist/workshop2/color_and_culture_files/frame.htm. (Accessed September 30, 2010).

Whorf, B.L. 1964. Language, thought, and reality. Boston: MIT Press.

Wickens, C.D., and Hollands, J.G. 1999. *Engineering Psychology and Human Performance,* 3rd edition. New York: Prentice Hall.

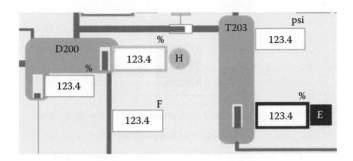

FIGURE 5.1 Example of using shape as a redundant code to color in depicting alarm conditions in an industrial schematic. Also note the use of gray tones for the plant schematic, with color reserved only for alarm conditions. (Reprinted with permission from Bullemer, Reising, Burns, Hajdukiewicz, and Andrzejewski, Effective Operator Display Design. Abnormal Situation Management Consortium, 2008. http://www.asmconsortium.org/)

FIGURE 5.2 The website of the Indian ICICI bank uses colors (**see color insert**) effectively to give users a positive affective experience. (From French, T., Minocha, S., and Smith, A. 2002. eFinance Localisation: an informal analysis of specific eCulture attractors in selected Indian and Taiwanese sites. In: Coronado, J., Day, D., Hall, B. (Eds.), Designing for Global Markets, Proceedings of IWIPS 2002, vol. 4. Products and Systems International, pp. 9–21.)

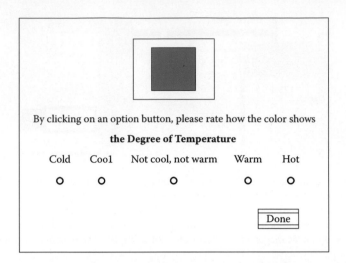

FIGURE 5.3 A sample color question. (From Liang, S.-F.M., Khalid, M., Taha, Z., and Plocher, T. In search of internationalized operator interface displays in process control: A comparison among Malaysian, Singaporean and Chinese. WWCS, 2004. With permission.)

FIGURE 11.8 Map of Suratganj drawn by villagers showing important features and facilities in relation to health-care facilities and functions.

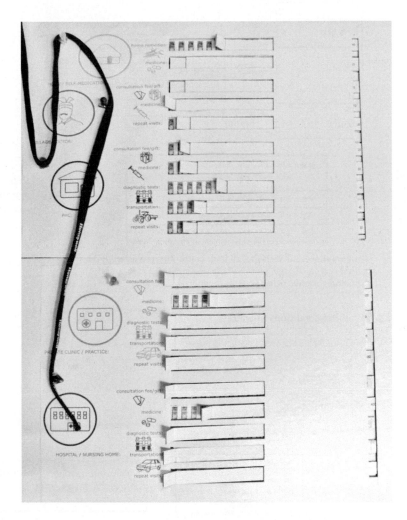

FIGURE 11.11 Participants used a Value Equation Tool for estimating their health-care expenditures.

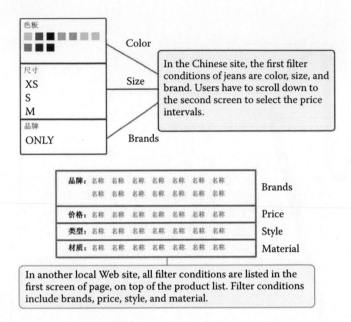

FIGURE 14.2 Filter conditions of jeans.

FIGURE 14.4 Product information with various alignments, fonts, character sizes, and colors.

6 Icons and Images

6.1 INTRODUCTION TO ICONS AND IMAGES

6.1.1 ICONS

Icons are small pictorial symbols used on graphic user interfaces (GUIs) to represent certain functions of the system. They are an essential part of any system with a GUI (i.e., windows, icons, menus, pointers [WIMP] interaction). The benefits of using well-designed icons are many. They can be an effective, nonverbal way to represent visual and spatial concepts. They foster immediate recognition by the user and aid in recall, reducing the user's reading time. They also help to efficiently use screen space. Horton (1994) stated that using icons can reduce the need for text translation. All of these can be of benefit to a product that is intended to be deployed globally (Horton, 1994).

Everyday objects, symbols, and gestures provide design inspiration and are commonly used in user interfaces, but they can be perceived differently in various parts of the world. For example, while a U.S. rural mailbox has been widely used to represent the concept of an e-mail account, it cannot be assumed that people from various cultures will perceive and recognize it as a mailbox. A common Japanese street mailbox looks like a U.S. trash can (Fernandes, 1995). The use of hand gestures in symbol or icon design is also problematic. The same hand gesture can be perceived differently, sometimes the opposite, by people with different cultural backgrounds. "Thumb up" is well known as "fine" or "good going" in North America and much of Europe, but it is perceived as insulting in Australia (Axtell, 1991). The "thumb up" gesture is also used in counting. For example, in Germany, a person uses the upright thumb to signal "one," and in Japan, the upright thumb is used to signal "five."

Other researchers emphasized that visual language can be problematic when it is used to communicate across cultures. The meaning of the image is highly negotiable in a cultural system (Fu, 2001), so icons can be effective in one culture but ineffective or downright offensive when used in another culture (Shirk and Smith, 1994). Some guidelines are provided below for designing effective icons for international use. The two case studies at the end of this chapter highlight some of the difficulties in designing icons that are understandable to a multicultural audience and the need to test them carefully before deploying them in a product.

6.1.2 IMAGES

The graphical images we portray in user interfaces, particularly Web pages, are often loaded with social and cultural connotations, statements of values, and attitudes (Marcus and Gould, 2000; Gould et al., 2000). For example, the selection of colors and artifacts such as national symbols provides a direct way to attract and satisfy website users in the target culture (Sheppard and Scholtz, 1999).

However, Gould (2001a, 2001b; Gould et al., 2000) points out that the graphics on a website often are much more than just symbols and potentially affect users in complex, subtle, and highly interactive ways. Gould suggests that users react to the social and cultural content in images by either identifying with the values and attitudes depicted or by rejecting them. The semantics of the image—the people and activities depicted, the image perspective relative to the viewer, the viewing angles, and the framing—are capable of evoking reactions of comfort or discomfort, satisfaction or dissatisfaction, pleasure or displeasure. These image semantics are what Kress and van Leeuwen (1996) refer to as the "visual grammar" of the image that draws the user into something close to a social interaction with the image. They write:

> The articulation and understanding of social meanings in images derives from the visual articulation of social meanings in face-to-face interaction, the spatial positions allocated to different kinds of social actors in interaction (whether they are seated or standing, side by side or facing each other frontally, etc.). In this sense the interactive dimension of images is the "'writing" of what is usually called "non-verbal communication," a "language" shared by producers and viewers alike. (Kress and van Leeuwen, 1996)

Simply put, the composition of the image creates social, affective ties between the viewer and the image that mirror real-life social dynamics between the viewer and the purveyor of the image (e.g., the customers, workers, and corporations, students and universities, citizens and governments, etc.). Gould (2001a, 2001b) interpreted Kress and van Leeuwen's (1996) work on visual grammar into the context of culture and cross-cultural design. She suggests that graphics and images should be composed in such a way as to match the social and cultural values and attitudes of the target culture. An image on a website that is poorly matched to the values and attitudes of the target culture runs the risk of ineffectively communicating its message. In the next section, we present some guidelines for designing images into the GUI or website.

6.2   GUIDELINES

6.2.1   MAKE SURE ICONS ARE HIGHLY RECOGNIZABLE TO THE TARGET USERS

6.2.1.2   Why?

A number of researchers have written about designing icons for specific cultures or for international use. Shen et al. (2007) reported on Chinese Web design with cultural icons. Cultural issues in designing international biometric symbols are described in Choong, Stanton, and Theofanos (2010). Pappachan and Ziefle (2008) discussed cultural influences on comprehensibility of icons. Kim and Lee (2005) reported on cultural differences in icon recognition on mobile phones. A series of international studies on telecommunications icons designed and tested by ETSI (European

Telecommunication Standards Institute) indicated that people from five Southeast Asian countries encountered difficulties in recognizing and using the symbols (Piamonte, Ohlsson, and Abeysekera, 1997; Piamonte, Abeysekera, and Ohlsson, 1999). Major software companies also present guidelines around this design issue (IBM, 2012b; Microsoft, 2012; Oracle, 2010).

6.2.1.3   How?

From these studies we conclude that a good icon for global use has the following properties:

- Mimics both the physical appearance and the function or action of the object it represents
- Clearly represents the state of the object if it is an object that can assume more than one state
- Uses only widely recognized conventions for color and shape
- Is not directional and can be used without rotation
- Uses a culturally neutral metaphor (if metaphor is used)
- Does not have embedded text characters

If evaluation and redesign fails to produce icons that are recognized by multiple target groups, then localizing the problematic icons to specific target cultures via locally meaningful metaphors might be necessary.

6.2.1.4   Example

Working at Motorola, Tham and Tan (1999) led two design teams, one in the United States and one in Singapore, to design the GUIs for two pager/personal digital assistant (PDA) products targeted for the Chinese market. Each team relied on images and symbology from their respective cultures. The icons designed by the Asian team used caricature or "cartoon" style in their icons. The American icons were more conventional computer icons. Results showed that users in the target market (China and Singapore) preferred the design by the Singapore team. Figure 6.1 illustrates the two designs.

6.2.2   WHEN DESIGNING ICONS, PROVIDE A COMBINATION OF TEXT AND PICTURE

6.2.2.1   Why?

Embedding text inside an icon is never a good idea. However, placing a text label in the user's language just outside the icon or providing a "mouse over" label can be an effective way to increase its meaning. Sometimes this practice is challenged by design constraints. For example, the decision to use a text label with each icon might be affected by the desire to avoid translation costs, the need to cater to a nonliterate target population, or limitations of screen space. Whatever the rationale for not using text labels, the decision may have consequences for cross-cultural usability.

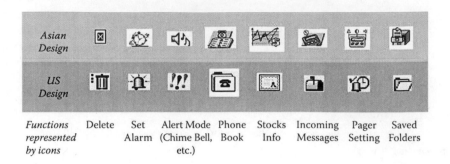

Functions represented by icons	Delete	Set Alarm	Alert Mode (Chime Bell, etc.)	Phone Book	Stocks Info	Incoming Messages	Pager Setting	Saved Folders

FIGURE 6.1 Two designs of graphic user interfaces (GUIs) for two pager/personal digital assistant (PDA) products targeted for the Chinese market. (From Tham, M.P., and Tan, K.C. 1999. Challenges in designing user interfaces for handheld communication devices: A case study. In Hans-Jörg Bullinger, Jürgen Ziegler, editors, *Human-Computer Interaction: Ergonomics and User Interfaces,* Proceedings of HCI International 99 (the 8th International Conference on Human-Computer Interaction), Munich, Germany, August 22–26, 1999, Vol 1, pp. 808–812. Mahwah, NJ: Lawrence Erlbaum, 1999. With permission.)

Choong and Salvendy (1998) examined the performance differences between the American and the Chinese users in recognizing icons presented in different modes (combined, text only, and picture only). The results indicated that a combined—text plus picture—presentation mode was at least as good as or better than the performance with either text alone or picture alone. However, when testing the alternatives, the study also found that American users performed better with a text-only presentation and Chinese users performed better with a picture-only presentation. This is not surprising, given the pictorial nature of the Chinese language versus the symbolic nature of English. Similar findings were reported that bimodal (text and picture) Chinese icons provide the best appropriateness and meaning for Chinese users (Kurniawan, Goonetilleke, and Shih, 2001).

6.2.2.2 How?

Whenever feasible, associate a text label in the local language with each icon. This ensures that both types of users—iconic and textual—will be comfortable with the presentation (IBM, 2012b).

6.2.3 Make Sure the Textual Components of Graphics Are Compatible with the Language(s) of the Target Users

6.2.3.1 Why?

Icons with embedded text will require translation and recompiling of the icon graphic in order to be transported to different cultures.

6.2.3.2 How?

The designer should minimize the use of words and abbreviations embedded in icons.

6.2.3.3   Example

Consider the example of Microsoft Word®. The "Font" icons use the alphabetic characters "A", "B", "I", and "U". These characters appear in the icons as an intrinsic part of the icon design. To transport these Font icons to a culture with a different language, the alphabetic component of the icon will need to be replaced with appropriate words in the language of the target users.

6.2.4   DESIGN GRAPHICS TO SUPPORT NATURAL READING AND SCANNING DIRECTION

6.2.4.1   Why?

The natural scanning direction, left to right or right to left, to which the user is accustomed influences his or her interpretation of the sequence of events in a complex graphic, the relative importance and value of objects in a graphic, and the recognition of text associated with the objects.

6.2.4.2   How?

Horton (1994) suggests certain steps that can be taken to avoid misperception or misinterpretation of graphics caused by differences in users' reading directionality. Those steps include

- Provide an indication (e.g., arrow) for the sequence of objects.
- Use a vertical sequence.
- Flip the graphics if appropriate.
- Design images that are symmetrical.

6.2.4.3   Example

The icons shown in Figure 6.2 and their meaning all have a strong left-to-right reading bias.

6.2.5   AVOID USING GRAPHICS WITH CULTURE-SPECIFIC METAPHORS AND ASSOCIATIONS

6.2.5.1   Why?

One of the most recommended techniques for user interface design is the use of metaphor. Metaphor enables users to associate their real-world knowledge and experience with a computer system function. Most metaphors use visual and conceptual representations of major user objects and their associated actions. A good metaphor

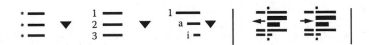

FIGURE 6.2   Examples of icons with left-to-right reading bias.

can help users connect what they do not know with what they do know, their real-world experiences.

That said, the use of metaphors in graphics can be problematic when the same metaphor must be meaningful to users in multiple target cultures. The associations made by users rely greatly on their ability to recognize the patterns of the objects presented. However, as pointed out by Fernandes (1995), the look and feel of the real world varies from place to place. People have internalized perceptions of what looks local and what looks foreign. Common everyday objects are not the same every-where in the world.

6.2.5.2   How?

When designing cross-cultural user interfaces, the use of metaphors becomes a challenge as the real world changes from culture to culture. Many companies choose to localize metaphors in their user interfaces by only redesigning the objects or translating the text in a certain metaphor. However, "translating" a metaphor is not sufficient; the entire metaphor will need to be reevaluated and possibly replaced to make the interface mapped to the target users' cultural experience (Evers, 1998).

Also, for global designs, avoid using graphical metaphors based on visual puns, verbal analogies, gestures and body parts, and religious and national symbols. If one must design for one specific locale, then make sure each graphic conveys its intended meaning in each target locale (IBM, 2012b; Microsoft, 2012).

6.2.5.3   Example

For example, while postal service may be a common concept around the world, the physical mailboxes are very different in color and shape. When presenting a U.S. rural mailbox, it cannot be assumed that people from various cultures will perceive and recognize it as a mailbox. Initial users of the Mac in the United Kingdom perceived the trash can as a mailbox (Fernandes, 1995). A common concept used in Web design is the concept of a "home" page that denotes the starting point of a website or a Web application. The concept of "home" is common around the world and known to most people. However, the graphical representation of a "home" varies greatly across different cultures. For example, some software uses a mouse in the icon representing the point-and-click device. In some languages, the name for the pointing device is not the same as that of a small rodent. This icon will then need to be completely redesigned for users with other languages who do not refer to the device as a "mouse."

6.2.6   Make Use of Appropriate Symbols, Images, Graphics, and Colors That Are Highly Recognized in the Target Culture to Excite and Please the User

6.2.6.1   Why?

Culture can be broken into four main components: values, rituals, heroes, and symbols (Hofstede, 1997). Fernandes (1995) points out that quite often a design will incorporate one or more of these components and, consciously or unconsciously, affect the user. Cultural images represent both a potential problem and a potential opportunity.

If the use of these cultural components, symbolically, in a user interface is poorly conceived, it can insult people's values and reduce acceptance. One example of the potential problems with symbols is hand gestures. The same hand gesture can mean different things, sometimes the opposite, for different cultures. For example, "thumb up" means "good" in the United States, but it is insulting in Australia.

On the other hand, the thoughtful use of cultural symbols for a specific cultural audience can evoke positive user response, attract the user to the product, and add to a satisfying experience.

6.2.6.2 How?

Before using symbols with cultural connotations, research them and assess the cross-cultural issues associated with using them.

6.2.6.3 Example

French et al. (2002) analyzed Indian and Taiwanese e-Finance websites. In the SinoPac Bank home page, shown in Figure 6.3, the "double-fish" sign can be seen to provide strong bonds. It serves as a visual reminder to local residents who are familiar with the physical decoration of the building, and conveys the message of "prosperity" (the fish)—something will always be left over each year after a Chinese

FIGURE 6.3   Use of "fish" symbols to reinforce message of wealth and enriching life.

FIGURE 6.4   National symbols, such as these on a famous multinational corporation, engage local users.

New Year. Other excellent examples of effective use of local cultural symbols can be found in the country websites of a famous multinational corporation. As shown in Figure 6.4, the corporation used famous buildings or city scenes such as Manhattan for the U.S. site and the Temple of Heaven in Beijing for the China site.

6.2.7   ENSURE THAT GRAPHICS REFLECT, OR AT LEAST DO NOT CONTRADICT, THE DOMINANT SOCIAL VALUES OF THE TARGET LOCALE FOR SOCIAL DISTANCE, POINT OF VIEW, DEGREE OF INVOLVEMENT, AND POWER

6.2.7.1   Why?

Whether we intend them to or not, the images and graphics we use in GUIs convey values and attitudes about human relationships (Gould, 2000a, 2000b; Gould et al., 2001). Every culture has a set of values and attitudes that surround face-to-face social interactions. Some cultures are characterized by hierarchies of authority that govern social interactions between the levels of society. Others emphasize equality in human interactions. Some value and encourage inclusiveness and group membership, while others value individualism. The sexes are considered equal in some cultures, while in others, males or females are dominant. The manner in which we depict people in images on the screen directly affects the user or viewer of the image. Just as people in different cultures have different attitudes toward social interactions, so too will people in different cultures have different affective or emotional reactions to the same image. They will be more or less comfortable with the image and more or less accepting of the graphical message.

6.2.7.2   How?

What exactly is it about an image of a person or persons on the screen that users respond to emotionally and with culturally determined attitudes and values? Gould and colleagues (2000a, 2000b, 2001) suggest that we consider three characteristics of the graphical image on the screen:

1. Social rank and gender of the people depicted and the activity in which they are engaged
2. Perceived or apparent distance of the user/viewer to the person in the image (manipulated by size of the image)
3. Position and orientation of the person depicted on the screen relative to the user/viewer (manipulated by viewing angle)

6.2.7.2.1   Social Rank and Gender

Consider the position in society of the people depicted in the image on the screen and what they are doing. The people in the image could be authority figures, carrying out their work of managing subordinates, making money, wielding political power, and so forth. Or, the image could depict people on the same social level as the user or viewer. Two university website home pages, Figure 6.5 from Taiwan and Figure 6.6 from the Netherlands, illustrate this. Taiwan is a culture that traditionally

FIGURE 6.5 Social rank and gender shown on homepage from National Taiwan University located in Taiwan.

FIGURE 6.6 Social rank and gender shown on Web from Utrecht University located in The Netherlands.

has recognized the inequality between the powerful and less powerful members of society and valued respect and deference to authority. Notice that the university website from Taiwan depicts university officials performing ceremonial duties, not students engaged in student activities. The images are well-tailored to the social attitudes of the Taiwanese viewer. In contrast is the Netherlands, which traditionally is characterized by very few social barriers between levels of society. So it is no surprise that the Dutch university website shows students instead of leaders, and the students are engaged in informal social behavior.

FIGURE 6.7   Perceived distance shown on homepage from National Tsing Hua University in Taiwan.

FIGURE 6.8   Perceived distance shown on website from Utrecht University in The Netherlands.

6.2.7.2.2   Perceived Distance

Relationships of unequal power or authority are conveyed also by the perception of physical distance between the viewer and the person in the image. An image of a person in authority at a significant distance from the viewer (e.g., relatively small image on the screen) conveys aloofness, the impossibility of social contact. In contrast, "close-up" images of people tend to reduce social distance, suggesting that social interaction is within reach and perhaps even welcomed. The two university websites illustrate this characteristic of perceived distance (Figures 6.7 and 6.8). The university officials in the university website from Taiwan are portrayed as a relatively small image on the screen, at a great apparent distance from the viewer. The images of students in the Dutch website are much larger and at a much closer perceived distance.

FIGURE 6.9 Image position and orientation shown on homepage from National Tsing Hua University in Taiwan.

6.2.7.2.3 Image Position and Orientation

All images of people on the screen have a specific orientation to the viewer in terms of two parameters: vertical angle and horizontal angle. Think of these as "camera angles." These visual angles have emotional connotations for the viewer which are culture based. Consider first the vertical angle of an image relative to the viewer. A high vertical angle, positioning the viewer above the person in the image, conveys social power to the viewer. A low vertical angle, positioning the viewer "below" the person in the image, conveys social power and authority to the person in the image. Eye-level vertical angles place the viewer on an equal social level with the person in the image. The Web page shown in Figure 6.9 illustrates.

Consider now the horizontal angle of the image to the viewer. Images that are positioned facing directly at the viewer and perhaps even making eye contact (Figure 6.10) draw the viewer into the group or activity depicted on the screen. In contrast, oblique image-to-viewer angles serve to isolate the viewer and exclude him or her from the group or activity depicted in the image. The latter effect can be used to further enhance the exclusiveness or position of the person in the image.

FIGURE 6.10 Image position and orientation shown on web from Utrecht University in The Netherlands.

Given all of the above, we must always remember that culture is not static. Due to social and political dynamics, it evolves with each generation of users. Hence, it is important to include some counterexamples of websites from Taiwan that defy the traditional values that one would expect based on theory. The website home pages shown in Figures 6.11 and 6.12 are in stark contrast to those illustrated above. These reinforce the importance of tracking and understanding cultural change in

FIGURE 6.11   Web page from Taiwan Tech uses images in a manner that defies traditional values for social rank and gender and perceived social distance.

FIGURE 6.12   Web page from National Taipei University of Technology also uses images of young people of both genders presented at eye level to the viewer to defy traditional Taiwanese values of social rank, gender, and perceived social distance.

your target markets, knowing who your users are, and designing graphics that will attract and satisfy them.

To summarize, if we understand the cultural values and attitudes of our user, we can manipulate these characteristics of a graphical image on the screen to make the user feel comfortable and satisfied with the interaction. Alternatively, if we use images carelessly we run the risk of creating a mismatch between the user's values and attitudes and the ones conveyed by the image on the screen. Offending the user or even making him or her feel uncomfortable reduces the chance that he or she will continue to browse the website to the desired conclusion.

Thus, for an internationalized design, use culturally neutral viewer-image arrangements.

For culturally localized designs,

- Understand the attitudes of the target culture toward authority, equality of the sexes, social distance, and individual versus group orientation. Understand that this may vary with the generation of users you are trying to reach. Some portions of the population in a particular culture may be more traditional in values than others. Know which portion of the population you are trying to reach and what they value.
- In your graphics, depict people and their surroundings in ways that make the viewer comfortable. Graphical dimensions to manipulate include horizontal angle, vertical angle, size of frame, size of image/apparent distance, gender, age, and inclusiveness.

6.3   CASE STUDIES

6.3.1   CASE STUDY: CROSS-CULTURAL STUDIES ON ICONS

Typically, icons have two elements: a figural image and a textual label. When designing icons for international users with different levels of comprehension, it is also necessary for designers to consider cultural issues. Theory suggests that Chinese have relatively lower verbal abilities and superior visual discrimination abilities compared with Americans (Bond, 1986). This poses an interesting research question about the design of icons for Asians and Americans.

6.3.1.1   Objective

Research by Choong and Salvendy (1998) and later Rau et al. (2003) compared American and Chinese users in their performance with icons that were strictly pictorial or combined text and pictures to simply text.

6.3.1.2   Method

Researchers designed a set of 20 icons to emulate commands for a word-processing application. American, Chinese Mainland, and Taiwanese participants performed recognition tasks using different presentation modes of icon displays (text mode, pictorial mode, and combined mode) by their native languages. In the test, participants were required to learn the set of icons and the necessary actions first. Then

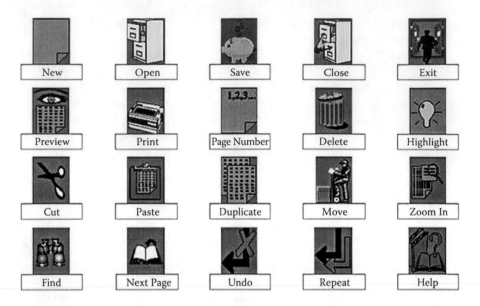

New	Open	Save	Close	Exit
Preview	Print	Page Number	Delete	Highlight
Cut	Paste	Duplicate	Move	Zoom In
Find	Next Page	Undo	Repeat	Help

FIGURE 6.13 Icons for American participants.

they needed to recognize the target icons according to the different tasks. There were three consecutive trials with 20 items in each task, which were the same for each participant. To avoid the other confounding variations among subjects other than cultural differences, only male undergraduate students majoring in engineering joined in the research. Their performance time and error rate were recorded for analysis. The testing system is illustrated in Figures 6.13 and 6.14 (Choong and Salvendy, 1998; Rau et al., 2003).

6.3.1.3 Results
In the first trial, American participants were significantly faster using icons in the combined pictorial-text mode than icons in the pictorial mode. Americans also were faster on the first trial using the text-only mode compared to the pictorial mode. Furthermore, American participants made the fewest errors when using the text-only mode. Participants from Mainland China were faster at the recognition task when using the combined mode or pictorial mode icons compared to the text mode. Taiwanese participants using combined modes were significantly faster than when they used the pictorial mode in the first test. Errors by participants from Mainland China and Taiwan using combined mode were significantly less than when they used the text mode (Choong and Salvendy, 1998; Rau, Choong, and Salvendy, 2003).

6.3.1.4 Conclusion
American participants performed better in a text-only mode and combined pictorial-text mode than in pictorial mode. Users from Mainland China and Taiwan performed better using icons in a combined pictorial-text mode than the text-only mode or pictorial-only mode. Participants from Mainland China also performed well with pictorial mode icons (Choong and Salvendy, 1998; Rau et al., 2003).

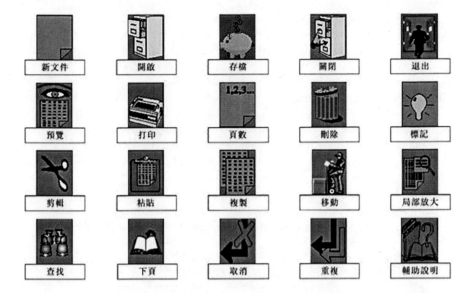

FIGURE 6.14   Icons for Chinese participants (Traditional Chinese).

6.3.1.5   Application

Designing icons that are pictorial but also associated with a text label will accommodate users from pictorial language traditions, such as Chinese, as well as users from more verbal cultures such as the United States.

6.3.2   CASE STUDY: DESIGN AND EVALUATION OF SYMBOLS FOR BIOMETRIC TECHNOLOGIES

6.3.2.1   Background

Biometric technologies are used to identify an individual based on biological and behavioral characteristics, such as face, iris, retina, fingerprint, vein, palm, voice, and signature. They can be used to prevent the illegal usage of an ATM, PC, or mobile phone, and are also useful in situations where user authentication is required, such as physical access of buildings or logical access of computers. With the development of biometric technologies and their increasing use, users may be unfamiliar with particular implementations, and they may not understand the local language in which instructions for use are described. It is important that the symbols used on biometric devices have consistent significance globally and are not offensive.

From past research, we know that there are differences among countries and cultures in cognitive styles, cognitive abilities, personality, and cultural patterns. People with different backgrounds, languages, cultures, religions, lifestyles, and education levels have varying perceptions and expectations of a given symbol. As many public biometric systems are used by people with different nationalities, a consistent international standard set of symbols would reduce the difficulty that the wider community experiences in finding and using biometric systems.

The National Institute of Standards and Technology (NIST), U.S. Department of Commerce, developed a set of icons, symbols, and pictograms for biometric technologies. The objective is to increase the user's understanding of how to use biometric recognition systems and to minimize the necessity of using any specific language (e.g., text) to guide users' interaction with these systems.

In the initial phase of a series of research studies, the NIST Biometrics Usability Group developed 24 symbols and pictograms for investigation. The first 20 symbols were intended to identify biometric modality and to represent concepts for directions and biometric sensor activity or feedback. Some concepts have multiple variants that were evaluated to identify the best symbol for the corresponding concept. Figure 6.15 shows the concepts represented.

Although each individual symbol was designed for a concept, it was intended that the symbols be combined to fully illustrate the biometric scanning processes. For example, in a customs or immigration environment, procedures constructed from the

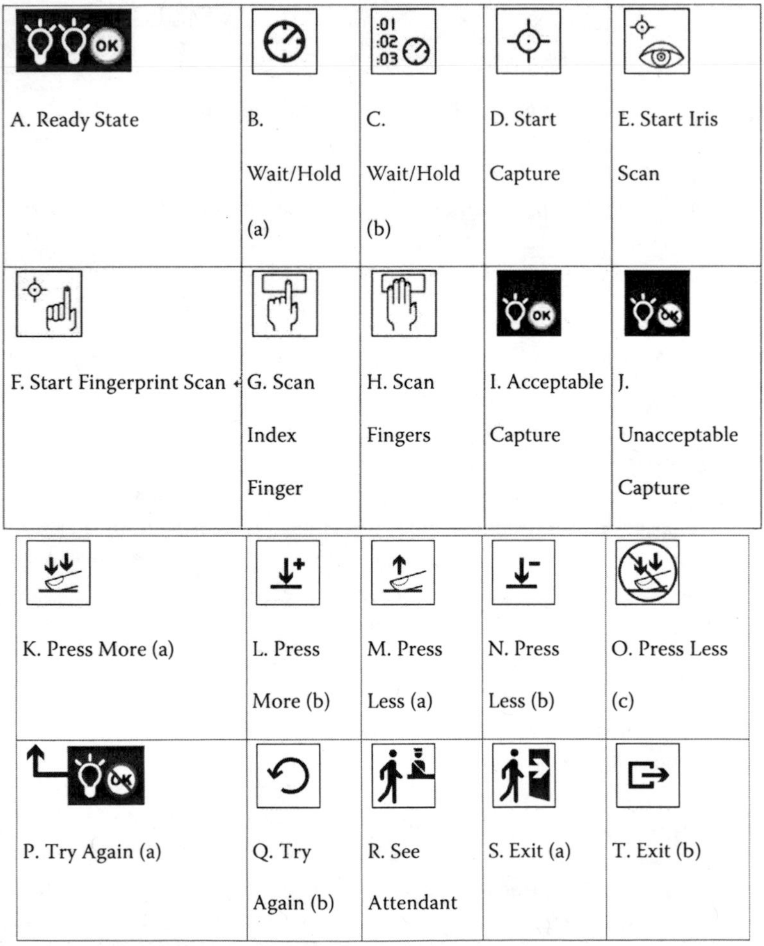

FIGURE 6.15 Biometric symbols under study, with intended meanings.

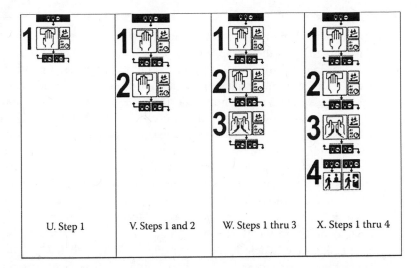

| U. Step 1 | V. Steps 1 and 2 | W. Steps 1 thru 3 | X. Steps 1 thru 4 |

FIGURE 6.16 Composite symbols representing steps in a fingerprinting procedure.

individual symbols can be presented as a series of posters while passengers are in the queue, or a series of transitional frames in a biometric booth. Figure 6.16 shows an example of a series of symbols constructed to represent a full capture procedure for "10-fingers," "print," "dual thumbs."

6.3.2.2 Objective

The NIST Biometrics Usability group conducted an international usability evaluation to investigate the comprehension and interpretation of biometric symbols by users with different cultural backgrounds and to suggest how to improve the designs. The usability evaluation was conducted first in the United States and then repeated in Japan, South Korea, Mainland China, and Taiwan independently.

6.3.2.3 Method

Researchers from NIST first conducted the usability evaluation in the United States and then in the Asian countries. Participants were recruited with the following criteria in mind:

- Mixed educational background
- Limited experience (shorter than 1 month) in United States, Canada, and Europe
- Equal number of male and female participants
- No age or occupational restrictions

The total number of participants in each country was as follows. In the first round of testing there were 15 participants from Mainland China, 12 participants from Japan, and 14 participants from South Korea. In the second round of testing, there were 100 participants from South Korea, 30 participants from Taiwan, and 15 participants from the United States. There are 186 participants in total.

6.3.2.4 Procedure

Before the usability test the English content and all the test materials were translated to the local language by the researchers from different countries. Participants performed the test individually. The test administrator greeted each participant and collected his or her demographic information. Then, the test administrator described the background of the study and the context in which the symbols would be used. There were two parts to the study: "Symbol Interpretation" and "Symbol and Meaning Matching." For the first part, the participants were shown the 20 proposed symbols (Figure 6.1), one symbol at a time, in random presentation order, and were asked to interpret the meaning of each symbol. The participants were then shown the four composite symbols in Figure 6.2, one at a time, following the sequence as in Figure 6.2, and were asked to interpret the meaning for each composite symbol. In the second part of the study, participants were asked to match a column of the 16 proposed symbols with a column of possible meanings. The task was to match the best meaning to each symbol. After part 1 and part 2, participants were asked about their opinion and subjective point of view of the symbols under study.

All the participants' interpretations from part 1 were compared with the original intended meanings by researchers and rated as one of the three levels: Correct Interpretation, Approximate Interpretation, and Incorrect Interpretation. A correct interpretation was defined as one that matched the researchers' intended meaning. For example, Figure 6.17 shows a symbol with the intended meaning of "start fingerprint." If a participant's response was, "to leave a fingerprint," which matches with the original meaning, then the participant's answer would be counted as a correct answer. An approximate interpretation refers to one that was not perfect but was close to the meaning intended by the researchers. As in the previous example, if the participant's reply was "touch here," it would be taken as an approximation to the correct "start fingerprint" interpretation. An incorrect interpretation refers to any interpretation of the symbol that was totally different from the intended meaning of the researchers. As in the previous example, something like "go that way" which does not even involve a finger or fingerprint, would have been scored as incorrect.

Part 2's matching data were calculated for the correct rate of each symbol.

6.3.2.5 Results

6.3.2.5.1 Recognition Tests

Symbols E—Iris Scan (Figure 6.18), and F—Fingerprint Scan (Figure 6.17):
Figure 6.17 shows that the intended meaning of F is "Start Fingerprint Scan," and Figure 6.18 shows that E is "Start Iris Scan." Yet many Korean participants interpreted symbol F as "pointing at something" and some interpreted symbol E as eyes but not iris (Kwon, Lee, and Choong, 2010). If the approximate interpretations are also considered, participants' interpretation for symbol E was closer (>50%) to the intended meaning for the United States, Japan, Korea, Mainland China, and Taiwan (Choong et al., 2010).

FIGURE 6.17 Start fingerprinting.   **FIGURE 6.18** E—Iris Scan.

FIGURE 6.19 Symbol D—Start Capture.

Symbol D—Start Capture (Figure 6.19): Most Korean, Taiwanese, and U.S. participants did not understand symbol D, illustrated in Figure 6.19, with interpretations such as "aiming or shooting" or "target," rather than the intended meaning of "Capture" (Choong et al., 2010; Ko, Choi, and Yi, 2010; Kwon et al., 2010; Liang and Hsu, 2010). The correct interpretation rate of symbol D was also lower than 20% in China and Japan where many participants interpreted the symbol as "scan area" and "press the target" (Choong et al., 2010).

Symbol B—Wait/Hold (Figure 6.20), and C—Wait/Hold (Figure 6.21): Few Korean participants identifed symbols B (shown in Figure 6.20) and C (as shown in Figure 6.21) correctly as "Wait/Hold," although symbol C got a relatively better result than symbol B (Ko et al., 2010; Kwon et al., 2010). This result was similar for participants in China and Taiwan who did not understand the intended meanings of the two symbols, although the symbol with additional information, C, was relatively better than symbol B (Liang and Hsu, 2010; Rau and Liu, 2010).

Symbols A—Ready State (as shown in Figure 6.22), I—Acceptable Capture (as shown in Figure 6.23), and J—Unacceptable Capture (as shown in Figure 6.24): Most participants did not interpret these correctly. They identified them as a "light," or "turning off/on a light," but not the intended meanings of "Ready State," "Acceptable," and "Unacceptable," respectively (Ko et al., 2010; Kwon et al., 2010; Rau and Liu, 2010). Participants in

FIGURE 6.20 Symbol B—Wait/Hold.   **FIGURE 6.21** Symbol C—Wait/Hold.

FIGURE 6.22   Symbols A—Ready State.

FIGURE 6.23   Symbol I—Acceptable Capture.

FIGURE 6.24   Symbols J—Unacceptable Capture

Taiwan felt confused about the symbols because they share common features, and thought symbols A and I were a pair, and symbols J and P were a pair (Liang and Hsu, 2010).

Symbols K—Press More (as shown in Figure 6.25), and L—Press More (as shown in Figure 6.26): Both symbols K and L were intended to represent the command "press more." The interpretations of symbol K tended to be "press" which is close to the intended meaning. The Korean interpretations of symbol L were either accurate or inaccurate and connected to Korean objects (e.g., "hospital") (Kwon et al., 2010). For China, Taiwan, the United States, and Japan, most participants interpreted both symbols correctly (Choong et al., 2010), with K performing better than L (Rau and Liu, 2010).

Symbols M—Press Less (as shown in Figure 6.27), N—Press Less (as shown in Figure 6.28), and O—Press Less (as shown in Figure 6.29): Symbols M, N, and O were different ways to depict the command "press less." Many participants from all test countries interpreted symbols M and O incorrectly. They interpreted the symbol O as "do not press," and the symbol M as "lift your finger." The correct rate of symbol N in the United States, Japan, Korea, China, and Taiwan was higher than symbols M and O (Choong et al., 2010; Ko et al., 2010; Kwon et al., 2010; Rau and Liu, 2010).

Symbols P—Try Again (as shown in Figure 6.30), and Q—Try Again (as shown in Figure 6.31): Not many Korean participants interpreted symbols P and Q correctly (Ko et al., 2010), although they identified symbol Q more readily than symbol P (Kwon et al., 2010). The result was opposite with participants in China, who understood symbol P better (Rau and Liu, 2010).

FIGURE 6.25   Symbol K—Press More.   **FIGURE 6.26**   Symbol L—Press More.

FIGURE 6.27   Symbol M—Press Less.

FIGURE 6.28   Symbol N—Press Less.

FIGURE 6.29   Symbol O—Press Less.

Symbols S—Exit (as shown in Figure 6.32), and T—Exit (as shown in Figure 6.33): For symbols S and T, both with the intended meaning of "exit," interpretations of Korean participants tended to split between the concepts of "enter" or "exit" (Ko et al., 2010), with relatively better interpretation of symbol T than symbol S (Kwon et al., 2010). However, for participants in China, the reverse was observed, symbol S having a higher interpretation rate than T (Rau and Liu, 2010).

Concrete Versus Abstract Symbols: Participants in China and Taiwan understood concrete symbols better than abstract symbols. From the results, symbols with concrete representation got higher correct interpretation rates than symbols with abstract representation (Rau and Liu, 2010). Yet there is one exception illustrated in Figure 6.27 and Figure 6.28: participants interpreted N better than M (Ko et al., 2010; Kwon et al., 2010; Rau and Liu, 2010), which may be due to the direction of the arrow in M. This confused participants and made them interpret M as "remove the finger" rather than "press less."

FIGURE 6.30   Symbol P—Try Again.

FIGURE 6.31   Symbol Q—Try Again.

FIGURE 6.32   Symbol S—Exit.

FIGURE 6.33   Symbol T—Exit.

Composite Symbols V to X: The composite symbols V to X, shown in Figure 6.16, represent steps in a full capture procedure. There were great differences between East Asians (e.g., China, Japan, Korea, and Taiwan) and Westerners (e.g., United States) in their ways of thinking about these composite symbols. The East Asian participants tended to find the relationships between the symbols during the interpretation. When the context was provided, users could understand the symbols better. As the symbols were put together to describe a context, East Asians participants comprehended the meanings more easily. Once subjects understood the first symbol (U), they interpreted the others correctly (Kwon et al., 2010; Rau and Liu, 2010). In contrast, U.S. participants paid more attention to the objects in the symbols and used more logical rules to think about them. They tended to describe the symbols in terms of the objects represented on the symbols and the objects' states (Choong et al., 2010).

6.3.2.5.2 Matching Test

Promising Symbols—E, G, I, J, L, N, and R (as shown in Figures 6.18, 6.17, 6.23, 6.24, 6.26, 6.28, and 6.32, respectively): Seven symbols, E, G, I, J, L, N, and R, had matching rates above 50% for participants in the United States, Japan, Korea, Mainland China, and Taiwan, and apparently were not affected by national culture (Choong et al., 2010; Ko et al., 2010; Kwon et al., 2010; Rau and Liu, 2010). We expect that when participants see these symbols in the intended context, they will be able to associate them with their intended meanings.

Symbols That Must Be Redesigned (as shown in Figures 6.17, 6.19, 6.20, 6.21, 6.22, 6.27, and 6.33): The matching rates of some symbols were below 50% in all the countries tested. Even with the textual cues provided, participants were still confused and could not relate to the intended meanings of these symbols. Symbols D, F, and M must be redesigned or replaced completely (Choong et al., 2010; Kwon et al., 2010). The results from Korea and Taiwan indicate that the symbols A, B, C, and T also should be redesigned for use in these countries (Kwon et al., 2010; Liang and Hsu, 2010).

6.4 CONCLUSION

In these symbol usability studies, the 24 symbols and pictograms created by the NIST were evaluated for use in biometric systems, mainly for fingerprint captures. Through these studies, problems with the symbols were explored and suggestions were gathered to improve their meaningfulness among different countries. It was shown that East Asian users understand action symbols better than feedback symbols and actually prefer to interpret feedback symbols into actions (Kwon et al., 2010; Rau and Liu, 2010). That is to say, East Asians would prefer that a symbol guide them directly to perform some action, rather than having to infer the needed action by themselves. Also, East Asian participants understood concrete and contextually related symbols better than abstract symbols. This phenomenon could be explained by the relational-contextual nature of reasoning and cognitive style of these cultures.

6.5 APPLICATION

These studies present empirical evidence of different understandings of symbols, pictograms, and icons among different countries. Through the analysis, it was possible to understand how people, from the United States, Japan, Korea, China, and Taiwan interpreted similarly or differently the biometric symbols and pictograms developed by the NIST Biometrics Usability group. These studies could be utilized as the basis for further research relevant to the designed symbols and pictograms for better understanding. Moreover, when developing symbols and pictograms for other biometric recognition systems, it would be possible to apply these results.

REFERENCES

Bond, M.H. 1986. *The Psychology of the Chinese People.* New York: Oxford University Press.

Choong, Y.Y., and Salvendy, G. 1998. Design of icons for use by Chinese in Mainland China. *Interacting with Computers*, 9: 417–430.

Choong, Y.Y., and Salvendy, G. 1999. Implications for design of computer interfaces for Chinese users in Mainland China. *International Journal of Human-Computer Interaction,* 11(1): 29–46.

Choong, Y.Y., Stanton, B., and Theofanos, M. 2010. Biometric symbol design for the public—Case studies in the United States and four Asian countries. Paper presented at 2010 AHFE International Conference, Miami, FL, July 17–20.

Evers, V. 1998. Cross-cultural understanding of metaphors in interface design. In *Attitudes toward Technology and Communication,* ed. C. Ess and F. Sudweeks, 2–11. London.

Fernandes, T. 1995. *Global Interface Design: A Guide to Designing International User Interfaces.* Chestnut Hill, MA: AP Professional.

French, T., Minocha, S., Smith, A. 2002. eFinance Localisation: an informal analysis of specific eCulture attractors in selected Indian and Taiwanese sites. In: Coronado, J., Day, D., Hall, B. (Eds.), Designing for Global Markets, Proceedings of IWIPS 2002, vol. 4. Products and Systems International, pp. 9–21.Austin, TX.

Fu, L. 2001. When West meets East: The intercultural communication challenge for graphic design in a global context. *Bulletin of the Fifth Asian Design Conference*, October 13–15, Seoul, Korea.

Gould, E.W. 2001a. More than content: Web graphics, cross cultural requirements, and a visual grammar. In *Proceedings of the Ninth International Conference on Human-Computer Interaction 2001 (HCI 2001).* 2: 506–509. New Orleans, LA.

Gould, E.W. 2001b. Using cross cultural theory to predict user preferences on the web. In *Proceedings of the Ninth International Conference on Human-Computer Interaction 2001 (HCI 2001).* 2: 546–547. New Orleans, LA.

Gould, E.W., Zakaria, N., and Yusof, S.A.M. 2000. Applying culture to website design: A comparison of Malaysian and U.S. Websites. In *Proceedings of the International Professional Communication Conference (IEEE Professional Communication Society)/ SIGDOC 2000,* 162–171. Cambridge, MA.

Horton, W. 1994. *The Icon Book: Visual Symbols for Computer Systems and Documentation.* New York: Wiley.

IBM. 2012b. Globalize your business: guidelines to design global solutions. At: http://www-01.ibm.com/software/globalization/guidelines/.

ITU-T Telecommunication Standardization Sector of ITU. 1994. Procedures for designing, evaluating and selecting symbols, pictograms and icons. ITU-T Recommendation F.910. International Telecommunication Union, Geneva.

Kim, J.H., and Lee, K.P. 2005. Cultural difference and mobile phone interface design: Icon rec-
ognition according to level of abstraction. Paper presented at the Seventh International
Conference on Human-Computer Interaction with Mobile Devices and Services,
Salzburg, Austria, September 19–22.

Ko, S.M., Choi, J.K., and Yi, Y.G. 2010. International study of NIST pictograms, icons and
symbols for use with biometric systems—The case of South Korea. Paper presented at
2010 AHFE International Conference, Miami, FL, July 17–20.

Kress, G., and van Leeuwen, T. 1996. *Reading Images: The Grammar of Visual Design.*
London: Routledge.

Kurniawan, S.H., Goonetilleke, R.S., and Shih, H.M. 2001. Involving Chinese users in analyzing the effects of languages and modalities on computer icons. In *Universal Access in
HCI: Towards an Information Society for All*, ed. C. Stephanidis, 491–495, Proceedings
of the HCI International Conference August 5–10, Vol. 3. Mahwah, NJ: LEA.

Kwon, Y.-B., Lee, Y., and Choong, Y.Y. 2010. An empirical study of Korean culture effects
in the usability of biometric symbols. Paper presented at 2010 AHFE International
Conference, Miami, FL, July 17–20.

Liang, S.F.M., and Hsu, P.H. 2010. Usability evaluation on icons of fingerprint scanning
system for Taiwanese users. Paper presented at 2010 AHFE International Conference,
Miami, FL, July 17–20.

Marcus, A., and Gould, E.M. 2000. Crosscurrents: Cultural dimensions and global Web user-
interface design. *Interactions*, 7: 32–46.

Microsoft. 2012. Globalization step-by-step. At: http://msdn.microsoft.com/en-us/goglobal/
bb688110.aspx.

Oracle. 2010. I18n in software design, architecture, and implementation. At: http://developers.
sun.com/dev/gadc/technicalpublications/articles/archi18n.html.

Pappachan, P., and Ziefle, M. 2008. Cultural influences on the comprehensibility of icons in
mobile-computer interaction. *Behaviour and Information Technology*, 27(4): 331–337.

Piamonte, D.P., Ohlsson, K., and Abeysekera, J.D.A. 1997. Evaluating telecom icons among
Asian countries. In *Advances in Human Factors/Ergonomics, 21A, Design of Computing
Systems: Cognitive Considerations,* ed. G. Salvendy, M.J. Smith, and R.J. Koubek, 169–
172. New York: Elsevier.

Piamonte, D.P.T., Abeysekera, J.D.A., and Ohlsson, K. 1999. Testing videophone graphical
symbols in Southeast Asia. In *Proceedings of the Eighth International Conference on
Human-Computer Interaction (HCI99)*, 793–797, Munich.

Rau, P.L.P., Choong, Y.Y., and Salvendy, G. 2003. Effectiveness of icons and textural design
in human computer interfaces: A cross culture study of Chinese in Mainland China and
Taiwan and Caucasians population. *Asian Journal of Ergonomics*, 4(2): 73–90.

Rau, P.L.P., and Liu, J. 2010. Design and evaluate biometric device symbols for Chinese.
Paper presented at 2010 AHFE International Conference, Miami, FL, July 17–20.

Shen, S.T., Prior, S.D., Chen, K.M., and You, M.L. 2007. Chinese Web browser design util-
ising cultural icons. In *Usability and Internationalization, Part II, HCII 2007*, ed. N.
Aykin, LNCS 4560: 249–258. Berlin, Heidelberg: Springer-Verlag.

Sheppard, C., and Scholtz, J. 1999. The effects of cultural markers on Web site use. Fifth
Conference on Human Factors and the Web (HFWeb'99), Gaithersburg, MD. June 3.
http://zing.ncsl.nist.gov/hfweb/proceedings/sheppard/.

Shirk, H.N., and Smith, H.T. 1994. Some issues influencing computer icon design.
TechnicalCommunication, Fourth Quarter, 680–689.

Tham, M.P., and Tan, K.C. 1999. Challenges in designing user interfaces for handheld communication devices: A case study. In Hans-Jörg Bullinger, Jürgen Ziegler, editors,
Human-Computer Interaction: Ergonomics and User Interfaces, Proceedings of HCI

International 99 (the 8th International Conference on Human-Computer Interaction), Munich, Germany, August 22–26, 1999, Vol. 1, pp. 808–812. Mahwah, NJ: Lawrence Erlbaum, 1999.

7 Presentation Formats and Layout

7.1 INTRODUCTION TO THE PROBLEM

7.1.1 PRESENTATION FORMATS

Formats are specific to different regions. For example, numeric values can be represented in different ways. Sometimes separators are not used, and different regions use different symbols for separators and with different formats. For the same number, "1,234.56" is used in North America; "1.234,56" is used in Germany, Holland, and Italy; and "1 234,56" is used in France and Sweden. Other format conventions that need to be taken into account include calendars, date and time formats, names and addresses, telephone numbers, and currency.

Designers need to be aware of different measurement conventions for different regions such as dimensions, weights, temperatures, and paper sizes. Adequate accommodations need to be provided so that the product uses the appropriate measurements for the target regions.

7.1.2 LAYOUT

The orientation of information presented on information appliances has been predominantly of Western styles in which users read left to right, horizontally. As mentioned in Chapter 4, there are languages with different orientations that could impose different reading styles on users, thus affecting their expectation of the information presented on a user interface and their performance. For example, designers should pay special attention to presentation of text for the Chinese language, including how text is rendered on the screen, text alignment, and justification.

Figure 7.1 illustrates the different presentations.

In terms of direction of navigation, five navigation models are possible: left oriented, right oriented, top oriented, bottom oriented, and center oriented. The appearance of navigation buttons can be text based or graphics based. Lo and Gong (2005) examined the navigation model of e-commerce websites in the United States and China. The results showed that the Chinese sites favor the top-oriented navigation model, while the United States equally preferred left-, top-, and center-oriented navigation models. It is not surprising that Chinese sites favor the top-oriented navigation model, because traditionally Chinese writings are read from top to bottom and right to left. As to the appearance of navigation buttons, it was found that U.S. sites favor GUI-based navigation buttons, while Chinese sites favor text-based navigation buttons. Figure 7.2 illustrates different layouts.

FIGURE 7.1 Examples of traditional Chinese works on E-book (upper) versus English books on E-book (Lower) illustrate different orientations of information presented.

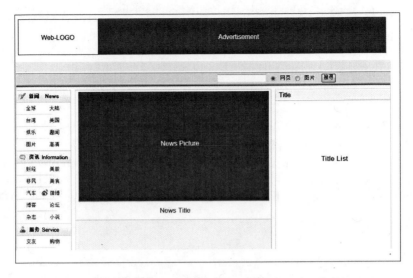

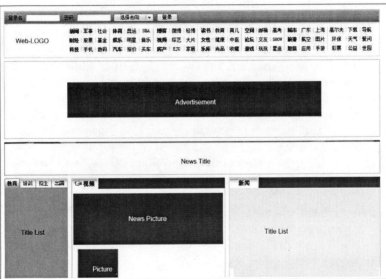

FIGURE 7.2 Examples of an Chinese Portal website of North American edition employs the left-oriented navigation model (upper) versus its Chinese edition employs the top-oriented & text-based navigation model (lower).

7.2 DESIGN GUIDELINES

There are existing guidelines (e.g., ISO 9241-14:1997(E)) about how to develop usable presentation and layout of GUI components, as well as the navigation among those components. The key is that the design has to match the user's task flow, experience, and expectation.

7.2.1 PROVIDE NATURAL LAYOUT ORIENTATION FOR INFORMATION TO BE SCANNED

7.2.1.1 Why?

Different cultures employ different format conventions and measurement systems that will affect the presentation of such information on the GUIs. Also, the arrangement of information on the screen affects how efficiently and comfortably people can scan, search for, and find the information. How efficiently people scan the screen and search for specific items of information is most related to the direction of their language.

7.2.1.2 How?

In Traditional Chinese, Korean, and Japanese languages, text is printed in columns, with breaks between the columns. It is read from top to bottom and horizontally (from right to left), following a typical "N" pattern. In cultures like Hong Kong and Taiwan, where the Traditional Chinese form is still used, people have a strong tendency to scan the screen "across the columns" or horizontally when searching for a specific piece of information. In contrast, Simplified Chinese can be printed either in rows or in columns, although it is mostly printed in rows and read from left to right and top to bottom (a "Z" pattern). As a result, people in Mainland China tend to scan the screen in a more vertical pattern. However, being accustomed to seeing both patterns of printed language, Mainland Chinese are able to adapt their search patterns based on the layout of the information (Goonetilleke, Lau, and Shih, 2002; Lau, Goonetilleke, and Shih, 2001). In a horizontal layout they search vertically or "row by row." In a vertical layout they search horizontally or "column by column."

Therefore, arrange the information on the screen in a way that is compatible (rows or columns) with the user's language so it can be scanned in a comfortable and efficient manner. This guideline applies to screens that display items of information with similar properties.

7.2.1.3 Example

This guideline applies to two-dimensional menus such as the item menus found on many online stores and thumbnail photo menus. It also applies any time you display a large field of data on the screen and the user's task is to search for a specific item. The guideline may also apply to searching Web pages, but there currently is no validation for that.

7.2.2 FOR MENU DESIGN, PROVIDE ORIENTATION COMPATIBLE WITH THE LANGUAGE BEING PRESENTED

7.2.2.1 Why?

It is common to have a menu bar placed horizontally in a user interface. However, as reported by Dong and Salvendy (1999), for Chinese users in Mainland China, users responded faster when presented with a vertical Chinese menu bar. The same group of users responded faster when presented with a horizontal English menu bar.

7.2.2.2 How?

The conventional guideline for displaying options in a pull-down menu is vertical orientation. Each item in the pull-down list of options is separated by a horizontal "line break." This is a natural way to present and read lists in languages such as English. However, some languages such as the Traditional Chinese used in Hong Kong and Taiwan, Korean, and Japanese are often written starting at the top right corner of the page. Text is written in columns from top to bottom. Natural breaks in the text are the spaces or "column breaks" between the columns. The most natural menu orientation for users of these vertically oriented languages is horizontal (Shih and Goonetilleke, 1997, 1998).

7.2.2.3 Example

Figures 7.3 and 7.4 illustrate the same "Find" function displayed with different orientations.

7.2.3 TEXT DIRECTION, LABELING, AND SCROLLING

7.2.3.1 Why?

Most guidelines use a left-to-right orientation or left justification for labeling text boxes, presenting text in text boxes, scrolling text within a text box, and presenting a series of control buttons. The guideline assumes that people read left to right. However, some languages such as Arabic are read from right to left.

FIGURE 7.3   "Find" functionality in horizontal direction

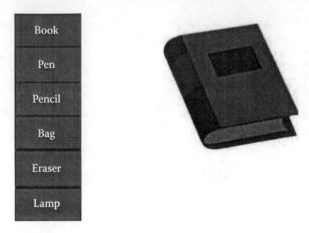

FIGURE 7.4   "Find" functionality in vertical direction

7.2.3.2 How?

To design a user interface that will accommodate a "right-to-left" language, you should adapt the display features that assume a certain text direction (Microsoft, 2012; IBM, 2012b). For languages with a right-to-left orientation,

1. Place text box label to the right of the box.
2. Right-justify the text within the box.
3. Scroll text in the box so it can be read from right to left.
4. Move the insertion point from right to left in front of the leading character.
5. If two or more buttons are used in the controls, and their order or frequency of use is important, then place them in a right-to-left order, with the most important one placed on the right.
6. Use right-to-left orientation to imply the order or sequence of items.
7. If an icon is associated with a line of text, position the icon consistent with reading direction (to the right of text item).

7.3 CASE STUDY: VISUAL SEARCH STRATEGIES AND EYE MOVEMENTS WHEN SEARCHING CHINESE CHARACTER SCREENS

7.3.1 BACKGROUND

How efficiently people scan the screen and search for specific items of information is highly related to the direction of their language. Content presented in a screen layout that is compatible with the directionality of the user's language will be scanned more efficiently. The classic experiment conducted by Goonetilleke et al. (2002) not only demonstrated these differences, but also showed that differences in scanning patterns exist even between Chinese-speaking users from different parts of Asia. This case study summarizes their experiment and its results.

7.3.2 METHODOLOGY

7.3.2.1 Experimental Design

The researchers hypothesized that

1. Population group (Mainland Chinese, Hong Kong Chinese, and nonnative Chinese reader), word complexity (high or low), and screen layout (horizontal, vertical, or uniform) and their interactions have a significant effect on search performance.
2. The main effects and interactions among the above variables will affect search strategy.

A 3 (Population) × 3 (Screen Layout) × 2 (Word Complexity) × 10 (Trials) factorial experiment was designed in which each participant performed in 60 trials (e.g., 10 trials for each Screen Layout × Word Complexity condition). The six conditions formed a Latin Square arrangement across which the testing order was balanced. Dependent variables were search time and percent correct.

7.3.2.2 Participants

The 18 paid participants included six Hong Kong Chinese, six Mainland Chinese with less than 1 year of experience living in Hong Kong, and six nonnative but fluent readers of Chinese characters. Cantonese was the first language of the Hong Kong Chinese participants. Chinese was not the first language of the nonnative group of participants. All were university students or lecturers. Also, all participants recognized and understood all 180 characters used in the experiment.

7.3.2.3 Experiment Paradigm

The researchers used a standard visual search paradigm in which a participant looks for one target item in a display with a number of distractors. The objective of the task was to search for and find a Chinese target character embedded in a full-screen search field of other Chinese characters. In this paradigm, the subject first was shown the target character. Then, when he or she was certain about the character he or she was supposed to search for, an experimental screen appeared in which the character was embedded in an array of other Chinese characters. In 9 of 10 trials, the target appeared randomly in 1 of 9 search areas of the screen, as shown in Figure 7.5c. In 1 trial out of 10, the target character was not present in the search field. The distractor characters were laid out in one of three arrays: horizontal rows, vertical columns, or a uniform distribution of characters. These arrays are shown in Figures 7.5a, 7.5b, and 7.5c. When the array appeared on the screen, the user's task was to search the array as quickly as possible and find the target character. Participants had 90 seconds to find the target. Search time and percent correct were recorded as measures. Correct responses were either "hits" or trials in which the participant correctly determined that the target was not present in the field. In addition, an eye tracker was used to record the user's visual search patterns for further analysis.

Because Chinese characters vary a great deal in complexity, two levels of character complexity were used in the experiment. High-complexity words had 16 to 18

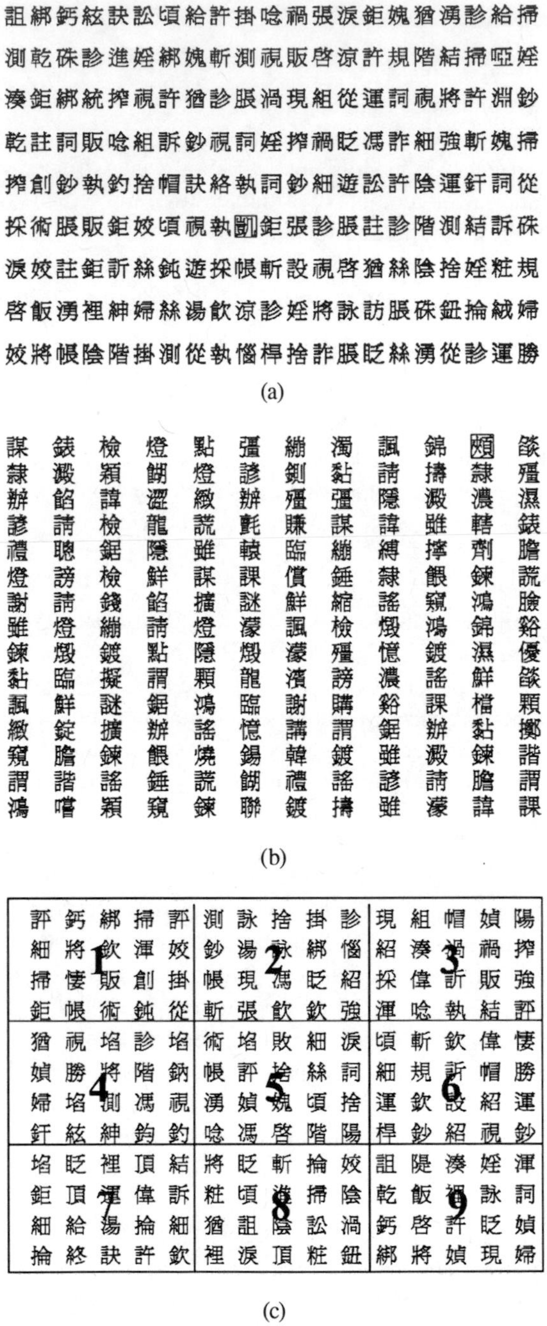

詛綁鈣絃訣訟頃給許掛唸禍張淚娓猶湧診給掃
測乾硃診進婭綁媿斬測視販啓涼許規階結掃啞婭
湊鉅綁統搾視許猶診脹渦現組從運詞視將許淵鈔
乾註詞販唸組訴鈔視詞婭搾禍眨馮詐細強斬媿掃
搾創鈔執釣捨帽訣絡執詞鈔細遊訟許陰運釺詞從
探術脹販鉅姣頃視執圓鉅張診脹註診階測結訴硃
淚姣註鉅訴絲鈍遊探帳斬設視猶絲陰捨婭粧規
啓飯湧裡紳婦絲湯飲涼診婭將詠訪脹硃鈕掄絨婦
姣將帳陰階掛測從執惱桿捨詐脹眨絲湧從診運勝

(a)

謀錢檢燈點彊繃濁諷錦頤餯
隸澱頹朗診剣黏請搗隸餯濕錢
辨餡譁澀辨癉彊隱濃轄濕膽諉
諉禮檢龍氈謀彊繃撑劑餯臉
禮燈錢隱輻賺鍾縛鍊濕餯谿
燈謝繃檢謎償檢餿窺鴻穎餯
謝雖燈錢燈濠鮮癉謠鴻錦擲
雖鍊煆鍍隱龍諷餯憶鍍模諧
黏臨鐔顆鴻臨諷濃鐔諧謂
諷鮮擴謠憶購謗餯課辨課
緻鍍謎餿錫謝購諉黏鍊譁
窺錠鍊錘翰禮謂鍍膽譁
謂膽諧窺聯鍍摶雖濠

(b)

鉼	鈣	綁	掃	鉼	測	詠	捨	掛	診	現	組	帽	娓	陽
細	將	飲	渾	姣	鈔	湯	綁	惱	診	紹	湊	禍	搾
掃	懷	販	創	掛	帳	馮	眨	紹	採	偉	訴	販	強
鉅	帳	術	鈍	從	斬	張	飲	欽	強	渾	唸	執	結	鉼
猶	視	垿	診	垿	術	垿	敗	細	淚	頃	斬	欽	偉	懷
娓	勝	將	階	鈉	帳	鉼	絲	詞	細	規	訴	帽	勝
婦	垿	測	馮	視	湧	垿	頃	捨	運	欽	設	紹	運
釺	絃	紳	鈞	釣	唸	馮	啓	階	陽	桿	鈔	紹	視	鈔
垿	眨	裡	頂	結	將	眨	斬	掄	姣	詛	隂	湊	婭	渾
鉅	頂	運	偉	訴	粧	頃	進	掃	陰	乾	飯	裡	詠	詞
細	給	湯	掄	細	猶	詛	陰	訟	渦	鈣	啓	許	眨	娓
掄	終	訣	許	欽	裡	淚	頂	粧	鈕	綁	將	娓	現	婦

(c)

FIGURE 7.5   (a) Example of a row layout. (b) Example of a column layout. (c) Example of a mix of uniform layout with the nine search areas also shown. Source: Goonetilleke, R.S., Lau, W.C., and Shih, H.M. 2002. Visual search strategies and eye movements when searching Chinese character screens. *International Journal of Human-Computer Studies*, 57: 447–468.

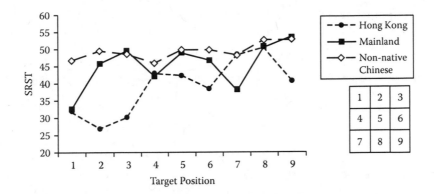

FIGURE 7.6 Different effects of target position on the search time of participants from different Chinese populations. Source: Goonetilleke, R.S., Lau, W.C., and Shih, H.M. 2002. *Visual search strategies and eye movements when searching Chinese character screens. International Journal of Human-Computer Studies*, 57: 447–468.

strokes, and low-complexity words had 10 to 12 strokes. Only Chinese characters with a left-right format were used because they can be rendered more clearly than top-down characters on a computer display. The size of the Chinese characters was 9 mm × 9 mm equivalent to a 1° visual angle. The target word was randomly chosen from a list of 36 words. The distractor characters were randomly chosen from a list of 100 characters, so each appeared one or two times on each screen.

7.3.3 Results

7.3.3.1 Errors
Analysis of variance (ANOVA) showed that while the nonnative participants had a significantly lower score on percent correct, the Hong Kong and Mainland Chinese groups did not differ from one another. Also, there was no effect of screen layout on errors.

7.3.3.2 Search Time
The researchers were concerned that starting position and search strategy would have a significant effect on search time. Therefore, in their analysis of search strategy, they treated Target Position (in search areas 1 through 9 on the screen) as an additional variable. To account for the nonnormality of the search time data as determined from a Q–Q plot normality test, a Square Root Search Time (SRST) transform was used. Using the transformed data, a four-way ANOVA was done for Population × Layout × Complexity × Target Position. The effects of Population and Target Position were significant, as well as the interaction between Population and Target Position.

Figure 7.6 plots search time for each of the three Chinese population groups across each of the nine search areas on the screen. It shows quite clearly that Mainland Chinese searched faster when the target was positioned in areas 1, 4, or 7, down the left-hand side of the screen. In contrast, Hong Kong Chinese searched most rapidly when targets appeared in areas 1, 2, or 3 across the top of the screen. Interestingly, the nonnative readers of Chinese appeared indifferent to target position.

No significant effect was found for the Word Complexity or Layout variables, although the authors noted a trend toward a significant word complexity effect. Also, the interactions were not significant, including the Population × Layout interaction.

7.3.3.3 Search Patterns

A manual review of the search pattern used on each trial revealed that some 90% of the searches were done systematically by the participant—that is, done by scanning row by row or column by column until the target was found. Typically, a search started at the top left or top right of the screen and continued with a search along that row or column. When finished with that initial row or column, the search moved over to the next one in order.

In order to better reveal and characterize each participant's vertical or horizontal search tendencies from the eye tracking data, the authors invented a new metric they dubbed the HV-Ratio (for Horizontal-Vertical). It is simply the sum of all the horizontal saccades (corrected for search field width) divided by the sum of all the vertical saccades (corrected for search field height). Participants who had a predominantly horizontal or row-by-row search pattern had an HV-Ratio greater than 1. Those for which a vertical or column-by-column search pattern dominated had an HV-Ratio less than 1. Those with a mixed or random search strategy, using equally, horizontal and vertical saccades, had HV scores around 1. Figures 7.7a, 7.7b, and 7.7c show examples of each search pattern including the recorded track of the eye, the derived search pattern, and the HV-Ratio score computed for that trial.

Figure 7.8 shows the effects of Population and Screen Layout on HV-Ratio scores. Statistical analysis of the HV-Ratio scores found that Hong Kong Chinese had a significantly higher HV-Ratio than both Mainland Chinese and nonnatives. That is, their search pattern was dominated by horizontal or row-by-row scanning. Mainland Chinese and the nonnatives were not different. More importantly, the interaction between Population and Layout was significant. Further analysis of the interaction showed that Layout did not significantly affect the HV-Ratio of Hong Kong Chinese and nonnatives. But Layout had a significant effect on Mainland Chinese who had their highest HV score for the row layout and their lowest for the column layout.

7.3.4 DISCUSSION

The results showed that Hong Kong Chinese have a tendency to search horizontally starting with the top row. They do this regardless of the directionality of the layout. The eye tracking data clearly showed that the horizontal searching was dominant in Hong Kong Chinese participants regardless of the screen layout. Some 90% of the eye movement data showed that searches started in the upper left corner of the screen. This combined with their strong horizontal search tendency suggests that Hong Kong Chinese will perform better if the targets of the search are placed in the top horizontal areas of the screen (e.g., areas 1, 2, and 3 in this experiment).

Mainland Chinese showed different search tendencies than the Hong Kong natives. They had the lowest search time for targets placed in the leftmost columns

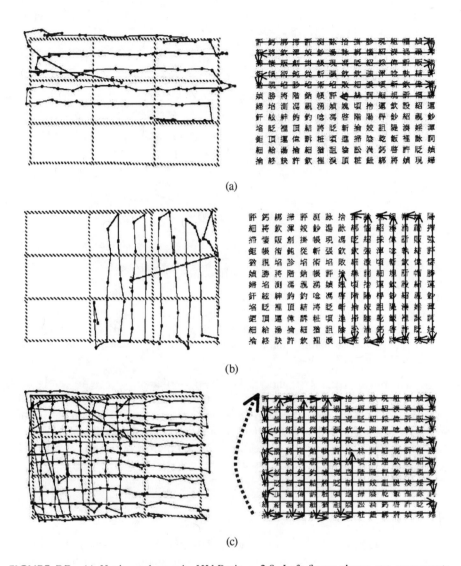

(a)

(b)

(c)

FIGURE 7.7 (a) Horizontal search. HV-Ratio = 2.8. Left figure shows eye movements. Right figure shows general pattern. (b) Vertical search. HV-Ratio = 0.17. Left figure shows eye movements. Right figure shows general pattern. (c) Combined search. HV Ratio= 1.0. User first searches screen using horizontal search, then switches to vertical search for second scan of the screen. Left figure show eye movements. Right figure shows general pattern. Source: Goonetilleke, R.S., Lau, W.C., and Shih, H.M. 2002. Visual search strategies and eye movements when searching Chinese character screens. *International Journal of Human-Computer Studies,* 57: 447–468.

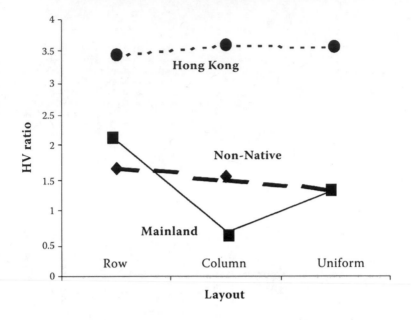

FIGURE 7.8   Effect of interaction between Population and Screen Layout on HV ratio. (From Goonetilleke, R.S., Lau, W.C., and Shih, H.M. 2002. Visual search strategies and eye movements when searching Chinese character screens. *International Journal of Human-Computer Studies*, 57: 447–468.)

of the screen (e.g., areas 1, 4, and 7). Area 1, in particular, was significantly faster than all other target areas. This likely shows that they prefer to start their searching in the upper leftmost area of the screen. However, once started, Mainland Chinese appear to flexibly adapt their search pattern to the layout of the screen. The HV-Ratio data showed that for row layouts, they used predominantly horizontal searches. For column layouts, they used vertical searches. Their HV-Ratios for uniform layouts were in between their high and their low HV-Ratio scores.

7.3.5   APPLICATION

Many screens require the user to systematically search for information of interest. Websites are a good example in which users have to search out the images, text, or hyperlinks that are important to them, amid all the other information displayed on the Web page. Good user interface design practice would have the most important or most popular items of information on a Web page placed in a location where the user would be most likely to see them. This case study showed that there are natural cultural differences in how people go about searching for information on a screen. For Chinese users, the common denominator appears to be a horizontal layout. Hong Kong Chinese perform best with a horizontal layout, and Mainland Chinese appear to readily adapt to it. Also, all the Chinese groups in this case study showed a strong tendency to begin their searching in the upper leftmost corner of the screen. From a design point of view, it would make sense to reserve this area on the screen or website for the most important items of information. Items of lesser importance or popularity are better placed in spaces lower and farther to the right on the screen.

REFERENCES

Dong, J., and Salvendy, G. 1999. Designing menus for the Chinese population horizontal or vertical? *Behaviour and Information Technology*, 18: 467–471.

Goonetilleke, R.S., Lau, W.C., and Shih, H.M. 2002. Visual search strategies and eye movements when searching Chinese character screens. *International Journal of Human-Computer Studies*, 57: 447–468.

IBM. 2012b. Globalize your business: guidelines to design global solutions. At: http://www-01.ibm.com/software/globalization/guidelines/.

Lau, P.W.-C., Goonetilleke, R.S., and Shih, H.M. 2001. Eye-scan patterns of Chinese when searching full screen menus. In *Universal Access in HCI: Towards an Information Society for All*, ed. C. Stephanidis, 367–371, *Proceedings of the HCI International Conference August 5–10*, Vol. 3. Mahwah, NJ: LEA.

Lo, B.W.N., and Gong, P. (2005). Cultural impact on the design of e-commerce websites: Part I– site format and layout. *Issues in Information Systems*, 6(2), 182-188.

Microsoft. 2012. Globalization step-by-step. At: http://msdn.microsoft.com/en-us/goglobal/bb688110.aspx.

Shih, H.M., and Goonetilleke, R.S. 1997. Do existing menu design guidelines work in Chinese? In *Advances in Human Factors/Ergonomics, 21A, Design of Computing Systems: Cognitive Considerations*, ed. G. Salvendy, M.J. Smith, and R. J. Koubek, 161–164. New York: Elsevier.

Shih, H.M., and Goonetilleke, R.S. 1998. Effectiveness of menu orientation in Chinese. *Human Factors*, 40(4): 569–576.

8 Information Organization and Representation, Navigation, and Hyperlinks

8.1 INTRODUCTION TO THE PROBLEM

8.1.1 INFORMATION ORGANIZATION

Differences in cognitive styles exist among people of different cultural backgrounds (Nisbett, 2003). Those differences have implications for how information should be represented and organized on graphic user interfaces (GUIs).

Oftentimes on a GUI, the information is organized into hierarchical structures. For example, menus display listings of choices or create a set of listings that guide a user from a series of general descriptors through increasingly specific categories until the lowest-level listing is reached. Design guidelines and best practices were developed for designing usable menus (e.g., Galitz, 1997; ISO 9241-14:1997(E); Weinschenk, Jamar, and Yeu, 1997). Those guidelines call for logical and meaningful organization of menu options. The menu options should be arranged into conventional or natural groups known to users and follow a logical order. The groupings should be logical, distinctive, meaningful, and mutually exclusive—but logical and meaningful to whom?

Menu structures for people in cultures with different cognitive styles need to accommodate the differences in how people organize information and represent the information accordingly. As Choong (1996) pointed out, when representing information on a GUI, Chinese users will benefit from a thematically organized information structure, whereas American users will benefit from a functionally organized structure.

Nawaz et al. (2007) reported similar results of cultural differences on the ways Chinese and Danish users grouped objects, functions, and concepts into categories. In the card sorting tasks, the Danish subjects preferred to highlight a category name by its physical attributes, whereas the Chinese subjects highlighted the category by identifying the relation between different entities. The Chinese subjects also utilized more thematic categories than the Danish subjects in the study by Nawaz et al. (2007).

Kim, Lee, and You (2007) reported similar findings with Korean and Dutch users interacting with menu structure designed for a mobile phone interface. The relational-grouping participants (Koreans) were more likely to select and prefer the thematically grouped menu, whereas taxonomic-grouping participants (Dutch) had the tendency to select and prefer the functionally grouped menu.

8.1.2 NAVIGATION

Navigation involves movement through the content and tools in an application. Examples of user interface (UI) elements that facilitate such movement include menus, windows, dialogue boxes, control panels, icons, and links. The different patterns of navigation are illustrated in Figure 8.1.

Marcus and Gould (2000) suggested that navigation will be impacted by cultures. Users from cultures who feel anxiety about uncertain or unknown matters prefer navigation schemes intended to prevent them from being lost (Marcus and Gould, 2000).

Kralisch, Berendt, and Eisend (2005) studied the impact of culture on website navigation behavior. In their study, they considered the impact on user behavior of Hofstede's (1980) long-term orientation and uncertainty avoidance and Hall's (1984) mono- and polychronicity. They collected behavioral data by sorting through records of navigation steps in the Web server log of a frequently used international multilingual website. The results demonstrated the impact of culture on website navigation behavior. Members of short-term–oriented cultures spent less time on visited pages than members of long-term–oriented cultures. In addition, more information was collected by members of high uncertainty avoidance countries than members of low uncertainty avoidance countries. Finally, monochronic cultures showed more linear navigation patterns than polychronic cultures, and vice versa. They suggested that for monochronic users, information should be placed in linear order, and links should emphasize hierarchical structure.

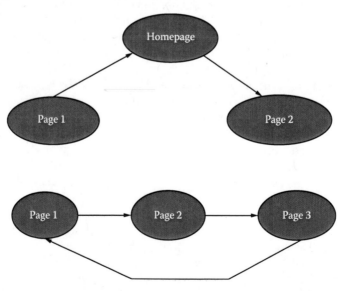

FIGURE 8.1 Examples of linear navigation sequence (upper) vs. nonlinear navigation sequence (lower)

8.2 DESIGN GUIDELINES

8.2.1 INFORMATION SHOULD BE ORGANIZED ACCORDING TO THE TARGET USER'S STYLE OF THINKING

8.2.1.1 Why?

People around the world hold different thinking styles, and the differences in thinking can affect their performance interacting with computers. Choong's (1996; Choong and Salvendy, 1999) research highlights that different cultures often focus on different attributes of the same items or objects. A website should be structured to reflect the target users' tasks and their view and organization of the information space to facilitate those tasks. Culture will influence what groups of items or functions should be placed together as pages and how links on a website or buttons and menus on the user interface should be labeled.

8.2.1.2 How?

Group items or functions and label lines on a website or buttons and menus in a user interface according to these categorizing preferences. Structure the information in an interactive system according to the target user's preferred way of organizing or categorizing information.

8.2.1.2.1 Identifying Categories and Structure in Information: Card Sorting

Card sorting is the best method to find out how your user prefers to categorize and label information and functions to be presented in a user interface or on a website. For a given UI design problem, it can be applied to two or more cultures to identify differences in how each prefers to organize the information.

Card sorting can be done by hand with paper note cards or with a computer program such as CardZort™. Start with a stack of blank paper or "electronic" cards. On each card, write one information item or function from the list of items or functions which needs to be presented in the user interface or website. One hundred cards is probably the maximum that you can deal with in one study. Recruit some representative users, paying special attention to the target culture of your product. Ask the users to sort the cards into piles that they think go together. There are no constraints on how many piles are allowed or the size of each pile. After the user sorts the cards, ask him or her to write down a name or label for each pile. These piles of closely related cards then serve as the basis for designing the lowest level of the information structure.

After completing the first sorting task, the users then are asked to try and group related piles of cards together and name them. This is the basis for designing a second level of information or functional organization in the user interface. You probably do not want to go to a third level.

Do this card sorting with at least 10 users from each target user group or culture.

Analyze the card sorting data by means of a cluster analysis using a standard statistical package or one specially designed for card sorting such as CardZort™. More advanced analyses of raw card sort data can be done with the University of Washington Card Sort Analyzer (Anderson, Anderson, and Deibel, 2004) . The latter provides capabilities for advanced manipulation and analysis of card sort data including:

- Quantifying the similarity between two sorts
- Characterizing the degree of similarity—average, maximum, and minimum—in a set of sorts sharing some attribute (e.g., culture)
- Finding clusters of sorts with a specified degree of similarity (e.g., "cliques")
- Analyzing a set of sorts for similarity to a theoretical archetype (such as thematic or functional)
- Searching on category names as a means to identify categories that have a similar semantic "gist"

A manual analysis of the data also is worthwhile, particularly to review the names and labels people have given different categories, and arrive at some consensus on this.

8.2.2 PROVIDE SEARCHING MECHANISMS AND PROCEDURAL SUPPORTS THAT ARE COMPATIBLE WITH THE USER'S TIME ORIENTATION

8.2.2.1 Why?

Researchers have indicated the significance of searching mechanisms for Web design. Morkes and Nielsen (1997) suggested that designers provide search mechanisms and structure information to facilitate focused navigation on all websites. They found that 79% participants scanned text, and only 16% read word for word. Nielsen (1997) found that only 10% of Web users would scroll a navigation page to see any links that were not visible in the initial display. Zhao (2002) studied the effect of time orientation on browsing performance and found that polychronic participants performed browsing tasks faster than neutral participants when participants were not familiar with the information architecture of the browsing materials.

8.2.2.2 How?

As a general rule, design for the monochronic user. Polychronic users will be able to adapt and cope with the monochronic design features. Following are some specific interaction design features to consider:

- *Provide Procedural Supports*: Monochronic users appreciate explicit procedures such as the following:
  - *Navigation*: For linear, procedure-oriented websites, such as those used for e-commerce transactions, list the explicit steps in the linear series of tasks required to complete the transaction. Number the steps in the sequence. Highlight the user's current place in the sequence of steps.
  - *Checklists*: For applications that require the user to find files of information from different sources, combine them and send them (e.g., a job application, customer proposal, expense report), and provide a checklist of the required items or actions.
  - *Error Messages*: Explain the error, but also provide a step-by-step procedure to fix it.
  - *Documentation and Help*: Provide specific, step-by-step procedures for performing tasks with the application.

- *Support Multiple Task Work*: Often the user will have several applications open and be working on several tasks during the same period of time.
  - *User-Specified Notifications*: The user can interact with only one task at any given moment. Changes or events may occur in his or her other tasks or applications but may not be detected by the monochronic user who remains focused on his or her current task. Therefore, notify the user of significant events or changes in these other ongoing tasks. Allow the user to specify the conditions under which he or she wants to be notified.
  - *Aids for Switching between Tasks*: Polychronic users will frequently switch their work from one task to another. Make it easy for a poly-chronic user to return to a task following a momentary distraction. Use the blinking cursor or highlighting to show the user where to resume his or her work.
  - *Display Active Applications and Tasks*: Display the active applications and tasks in a prominent place on the screen.
- *Information Structure*: A shallow information structure in a website requires less linking. Monochronic people, who tend to be relatively slow in their Web browsing, will appreciate this.

8.2.3 PROVIDE BOTH A SEARCH ENGINE AND A WEB DIRECTORY TO SUPPORT DIFFERENT NEEDS OF USERS

8.2.3.1 Why?

Users with different cultural backgrounds may have different needs for searching mechanisms. Most websites have two types of search mechanisms built in: Web directories and search engines. Fang and Rau (2003) examined the effects of cultural differences between the Chinese and the Americans on the perceived usability and search performance of Web portal sites. Chinese participants tended to use keyword search to start a task. If that failed after one or more trials, they would then try to browse the categories to complete the task. American participants tended to browse categories in the beginning of a task. They might use keyword search to supplement category search. Some Chinese participants reported that they tended to use keyword search first when the search task was difficult. Choosing keywords was relatively easier than going through several levels of categories for difficult search tasks for some Chinese participants. American participants were likely to use keyword search when they believed that they knew exactly what they were looking for.

8.2.4 PROVIDE POSSIBLE OUTCOMES AND RESULTS OF OPERATIONS AS MUCH AS POSSIBLE FOR ASIAN USERS OR USERS IN HIGH-UNCERTAINTY-AVOIDANCE CULTURES

8.2.4.1 Why?

For high-uncertainty-avoidance cultures emphasize simplicity in information and clear mental models. This helps users avoid disorientation and redundant cues (Marcus et al., 2000). Asian countries and some countries in the Middle East follow high power distance in daily life so that probabilities and "what-ifs" associated with different procedural options are recommended (Plocher, Gang, and Krishnan, 1999).

8.2.5   PROVIDE EXTRA NAVIGATIONAL AIDS FOR JAPANESE, ARABIC, AND MEDITERRANEAN USERS OR USERS IN HIGH-CONTEXT COMMUNICATION STYLE

8.2.5.1   Why?

Cyr and Trevor-Smith (2004) conducted an empirical comparison of German, Japanese, and U.S. website characteristics. They found different preferences for navigation and search capabilities in these different cultures. Japanese sites were twice as likely to use symbolic navigation tools as were the German or American sites (Cyr and Trevor-Smith, 2004). Preferences for vertical and horizontal menus were statistically significant. German and Japanese sites used a "return to home" button twice as much as the U.S. sites. As to the type of hyperlinks used, the results found that the number of external links and the functionality of links differed across cultures. External links were used in almost all Japanese sites, compared to only two-thirds of U.S. and German sites. The Japanese use symbols for links significantly more than do German and U.S. sites.

Rau and Liang (2003a, 2003b) pointed to well-designed navigational supports to combat the tendency toward disorientation of users in high-context cultures. Rau and Liang (2003a) used a survey designed by Plocher et al. (2001) to classify Web users as either high or low context on Hall's communication-style dimension. The results showed that high-context people browsed information faster and required fewer links to find information than did low-context users. However, high-context users had a greater tendency to become disoriented, and lost their sense of location and direction in hypertext. Low-context users were slower to browse information and linked more pages but were less inclined to get lost. In another study, Rau and Liang (2003b) investigated the effects of communication style on user performance in browsing a Web-based service. The results showed that participants with high-context communication style were more disoriented during browsing than were those with low-context communication style.

8.2.5.2   How?

- Provide well-designed navigation supports for users to understand where they are in their browsing or e-transaction task and how to move forward and backward. This will reduce the tendency toward disorientation among users from high-context cultures, and low-context users will not mind. Some effective navigation supports to reduce disorientation in the high-context user include the following:
    1. Identify your site on all your pages by placing the same name or logo in the same place on the page.
    2. Make your identifier on each page a hyperlink that always goes back to the home page.

3. For hierarchical sites, show the location of each page relative to the site structure. List all the levels of hierarchy above the current location and make the name of each level a link to its main content page.

4. For linear, more procedure-oriented sites, list the explicit steps in the linear series of tasks required to complete the transaction. Number the steps in the sequence. Highlight the user's current place in the sequence of steps. Avoid the use of "Next" buttons in linear sites.

5. Use a clear main headline on each page.

6. Use the page title in the HTML header definition to generate a meaningful name for each individual page so the user can locate it easily in his or her bookmark list.

7. Use standard conventions of underlined text for embedded links, blue for unvisited links, and purple for visited links.

8. Avoid using pull-down menus or graphics for links because they do not behave in a similar way to underlined text (e.g., they do not turn purple if they are linked to pages that users have already visited).

9. Make sure structural links (links that point to another level in the site structure) are consistent from page to page. Also make sure each structural link states the name of the level to which it points ("Go Up to Home Security Products") rather than some generic relationship ("One Level Up").

- Support low-context users with a help and documentation system that provides background information on features or functions as well as explicit procedures on how to perform a task with it.

8.3   CASE STUDY 1: CULTURALLY ADAPTED MOBILE PHONE INTERFACE DESIGN

8.3.1   BACKGROUND

Kim and Lee (2007) and Kim, Lee, and You (2007) studied the influence of cognitive style on mobile phone interface design for Korea and the Netherlands based on the theoretical notion that Eastern people tend to use a holistic thought process and Western people an analytical thought process.

8.3.2   OBJECTIVE

This study aimed to illustrate how cognitive styles differ across certain cultures and can influence preferred information structure and sequence in the mobile phone interface.

8.3.3   METHOD

As the related studies indicate, one can predict that user performance with and favorable attitudes toward a user interface will be enhanced when the interface design is compatible with the user's cognitive style. To grasp which cognitive style has a correlation with which element in a mobile phone interface, the user interface elements of a mobile phone were conceptually divided into three different layers: representation, menu structure, and interaction flow. An experiment was designed to understand how analytic-holistic characteristics of cognitive style differentially affected each layer.

8.3.3.1 Prototype Test

A mobile phone prototype was created, the main screen of which consisted of six menus (Call history, Messaging, Phonebook, Sound, Display, and Settings). Setting contents was possible by using two "cognitively thematic" menus, Sound and Display, or the "cognitively functional" Settings menu. For example, as shown in Figure 8.2, to set or change the wallpaper on the mobile phone using a more thematic approach, participants started from the Display menu, then went directly to My Pictures to select a certain picture, and then clicked Set as Wallpaper from among the options in a context menu that popped up from the right side at the bottom. Alternatively, participants were able to change the wallpaper using a more functional approach by accessing the Settings menu, entering Wallpaper in Display, and then finally selecting one picture from among a list of pictures.

Participants were asked to finish two tasks: change the ringtone and change the wallpaper to a specific example for each task by entering one of the six menus. The menu that they initially chose to use for the tasks was noted. After completing the two tasks, they were asked to perform the same tasks again but this time using the alternative method. In this manner, subjects were forced to experience and compare two different ways (thematic approach versus functional approach) of changing the settings and indicate which approach they preferred.

8.3.3.2 Cognitive-Style Test

The cognitive-style test was intended to discover whether an individual's cognitive style was taxonomic or relational. Twenty-six sets of images were selected as stimuli. One target picture and two alternative pictures were presented together, and participants were asked to select, as quickly as possible, the one alternative that best matched the given target picture. The two alternatives belonged to the same taxonomy as the target picture, and one shared a relationship with the target picture. For example, a picture of a mouse (or a rat) was presented as a target picture, and a squirrel and a piece of cheese were presented as alternatives. A mouse and a squirrel are both animals and so were considered to be in the same taxonomic category. However, a mouse eats cheese. Thus the mouse and cheese share a relationship. The pictures remained on the screen until participants made their choices. Three sets were given as warm-ups prior to the actual trials in order to help participants understand how to respond. The other 23 sets were presented in sequence.

8.3.4 Results

8.3.4.1 Categorization Style

On the cognitive-style test, the percentage of each response type (relational grouping or taxonomical grouping) was counted to yield a relative index of individual

FIGURE 8.2 Interaction sequence to change screen wallpaper. Top, left to right, shows an approach catering to a cognitively functional user; bottom, left to right, shows a more thematic approach. (From Kim, J.H., Lee, K.P., and You, I.K. 2007. Correlation between cognitive style and structure and flow in mobile phone interface: Comparing performance and preference of Korean and Dutch users. In *Usability and Internationalization*, ed. N. Aykin, Part 1, HCII 2007, LNCS 4559, 531–540. Heidelberg: Springer-Verlag. With permission.)

categorization style (100: strong taxonomic tendency ~ 1: strong relational tendency). The Korean group (M = 35.73, SD = 26.45) had a greater relational tendency than the Dutch group (M = 42.32, SD = 30.70), but the difference was not significant [$F(1,56)$ = 0.74, p = .39].

8.3.4.2 Initial Menu Choice

Over 70% of both national groups initially set the ringtone by using a Sound menu. The results indicated that there were no significant differences in national groups in the initial choice of menus between the two groups (p = 1). For the task of setting the wallpaper, most of the participants changed the wallpaper by using the Display menu, with no significant differences between the two national groups (p = .52). These results show that there were no cultural differences inherent in the menus that subjects initially chose as they performed the tasks.

8.3.4.3 Preferred Menus

After experiencing both the thematic and functional approaches to setting sounds and wallpaper, subjects were asked for their preference. For changing the ringtone, 53% of Dutch participants (n = 16) preferred the Setting menu, and 77% of the Korean participants (n = 23) preferred the Sound menu, a significant difference between the groups (p = .03). When changing the wallpaper, 53% of the Dutch participants (n = 16) preferred the Setting menu, and 73% of the Korean participants (n = 22) preferred the Display menu. This difference was not statistically significant

($p = .06$). However, this was closely related to the national tendencies found in the sound-setting task.

8.3.4.4   Correlation between Categorization Style and Preferred Menu

National group aside, categorization tendency was found to influence menu selection. The categorization tendency was different ($F(1,56) = 5.05$, $p = .03$) between a group selecting the Setting menu ($M = 53.35$, SD = 34.46) and the group selecting the Sound menu ($M = 34.26$, SD = 25.02) to change the ringtone. The group that selected/preferred the Setting menu had a tendency to be more taxonomic than the group that selected/preferred the Sound or Display menu in both tasks.

8.3.4.5   Impact of Prior Experience

As a result of Fisher's exact test, no correlation was found between the ways for changing the ringtone in their current mobile phones and their selected menu during the test ($p = 1$). Additionally, there was no difference between the ways for setting the wallpaper in their current mobile phones and the ways they performed on the test ($p = .73$).

8.3.5   DISCUSSION

Korean participants preferred a thematically grouped menu, and Dutch participants preferred a functionally grouped menu. The categorization tendency of the Korean group was found to be more relational compared to the Dutch group, but the tendency was not statistically significant. The sample size was not large enough to be generalized down to a collective cognitive style. With the small sample, individual cognitive styles were more apparent than collective cognitive styles. However, such collective cognitive styles have been shown by a number of cultural psychological studies (Nisbett, 2003). Thus, the correlation between individual cognitive style and menu structure found in this research can feasibly apply to a cultural level under the assumption that East Asians tend to make more relational groupings compared to Westerners.

8.3.6   CONCLUSION

This study hypothesized a correlation between cognitive style and menu structure in a mobile phone interface. Individual differences in categorization were significantly correlated with type of menu structure. In other words, participants selected and preferred the interface adapted to their cognitive style. Thus, the hypothesis was supported; it has substantial implications in designing a culturally adapted interface.

8.3.7   APPLICATION

This study shows the possibility of a cognitively adapted interface that uses the connection between a cognition style and the interface architecture. This offers the possibility of providing more logical and quicker access to any command or option on the mobile device and hence, a more pleasant user experience. Current structure and flow of the interface appear universal across cultural areas. However, the findings in this research may be helpful if designing an interface suitable to each cultural area on the basis of the fact that the cognitive styles of East Asians and Westerners differ from one another. Given that mobile phones have a limited number of menus due to their small screen size, it is necessary to organize the limited number of main menus appropriately in order to offer logical and quick access to any command or option. The findings in this study suggest that menus can be organized in different ways depending on users' cognitive styles.

8.4 CASE STUDY 2: CULTURAL EFFECTS OF KNOWLEDGE REPRESENTATION AND INFORMATION STRUCTURE IN A USER INTERFACE FOR ONLINE SHOPPING

8.4.1 BACKGROUND

The cognitive style of Americans is analytical– categorical (functional). As a result, Americans have a tendency to classify objects on the basis of functions or attributes that the objects have in common (Chiu, 1972, Nisbett, 2003). In contrast, Chinese people have a relational–contextual cognitive style (thematic). They tend to classify objects on the basis of the thematic relationships between the objects, rather than common attributes or functions (Chiu, 1972; Nisbett, 2003). These theoretical observations lead Choong and Salvendy (1999) and later, Rau, Choong, and Salvendy (2004) to speculate that these cultural differences in object classification style would have a significant effect on the structure of information preferred by Chinese users of a computer application compared to Americans.

A second major difference between the analytical-categorical and the relational-thematic styles of thought is how subjectivity is treated. The American way of thinking tends to be analytic, abstract, and imaginative—beyond the realm of the immediately apprehended. The Chinese way of thinking tends to be synthetic, concrete, and within the periphery of the visible world (Chiu, 1972; Lin, 1939; Northrop, 1946; Yang, 1986; Yang et al., 1963). The analytical style separates subjective experience from the inductive process that leads to an objective reality. The relational style of thinking rests heavily on experience and fails to separate the experiencing person from objective facts, figures, or concepts (Stewart and Bennett, 1991). These differences in dealing with subjectivity lead Choong and Salvendy (1999) and Rau, Choong, and Salvendy (2004) to hypothesize that Chinese users would benefit from a concrete representation of the information in an application to help them develop accurate mental models and perform the interaction tasks properly and efficiently. They speculated that, in contrast, Americans would perform without problems using a more abstract, text-based, representation of the system.

8.4.2 OBJECTIVES

This case study summarizes experiments conducted by Choong and Salvendy (1999) and Rau, Choong, and Salvendy (2004) to explore the two propositions laid out above. One, Chinese would perform better when interacting with information structured thematically rather than functionally and, after interacting with such a UI, would be able to recall more of the information they had seen. Two, Chinese would benefit from more a concrete, in this case, graphical, representation of the information, rather than an abstract, text-based one.

8.4.3 METHOD

8.4.3.1 Participants

Forty American participants and 40 Chinese participants from Mainland China were engaged in the first experiment. The American participants were Caucasian undergraduate students from Purdue University, ages 18 to 25, with mean computer experience of 8 years. The Mainland Chinese participants lived in Mainland China and had never been to the United States. They were undergraduate students from Beijing University, ages 18 to 23, with 1 year of computer experience. The second experiment had 40 Chinese participants from Taiwan. They were male undergraduate students from Yuan Christian University, ages 18 to 23, with 3.4 years of computer experience.

8.4.3.2 Design of the Experiments

The two experiments presented here as a case study considered two independent variables, Knowledge Representation and Information Structure, in a 2 × 2 (Knowledge Representation × Information Structure) factorial design. The Knowledge Representation variable had two conditions related to preferences for processing information, 1) Abstract and 2) Concrete. The Information Structure variable had two conditions related to how information was categorized in the system, 1) Functional and 2) Thematic.

Four different user interfaces were designed for the experiments: 1) abstract with a functional information structure, 2) abstract with a thematic information structure, 3) concrete with a functional information structure, and 4) concrete with a thematic information structure. A simulated online shopping web site was created to provide a task environment in which subjects could carry out information searching, e.g. online shopping, tasks. The abstract representation of the system, in both information structure conditions, was a hierarchical plain text description of the information in the shopping system. It consisted of categories, subcategories, and items. The concrete representation of the system, for both information structure conditions, was a graphical department store metaphor. The store had four floors representing the four top-level categories of items. Under each floor/category were subcategories, and finally, at the third level, items in the store. Figure 8.3 shows, side by side, English language examples of the user interface designed to appeal to "concrete" users and the one designed to appeal to "abstract" users. In the experiments, all text was translated to the appropriate local language of the participant, English for Americans, Simplified Chinese for Chinese subjects from Mainland China, and Traditional Chinese for Chinese subjects from Taiwan.

Figure 8.4 shows the information structure designed for users with a functional mental model of the information in the system. Figure 8.5 shows the information

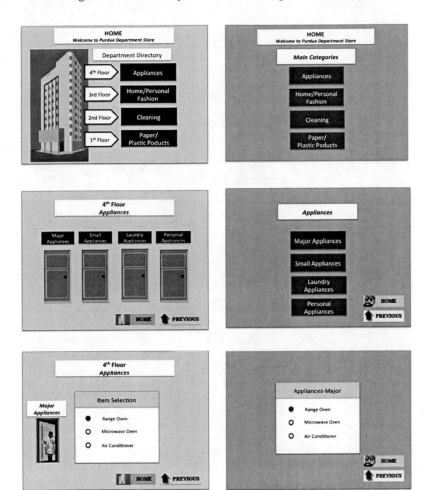

FIGURE 8.3   Examples of the user interfaces used in the experiment. Left side, top to bottom used a concrete knowledge representation scheme. Right side, top to bottom, used an abstract knowledge representation scheme.

structure designed for users with a more thematic model of the information. The functional style of organizing the information focused more on the physical and mechanical attributes of the objects/items in the store and commonalities in those attributes. Items with similar physical or mechanical attributes tended to be grouped together in the online shopping system. For example, all electric appliances were grouped together. In contrast, the thematic style of organizing the information focused more on the holistic relationships between objects/items, particularly the typical environment in which they are used. So, for example, all items typically used in a kitchen were grouped together.

Subjects were randomly assigned to one of the four user interfaces. A brief practice session was conducted to help the participant understand the operation of the system and the tasks to be performed. Following the practice, each participant

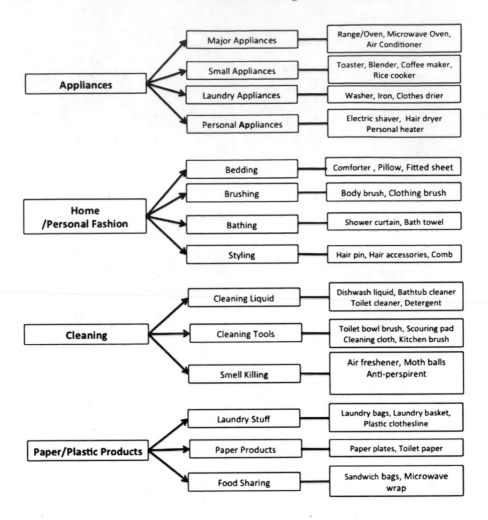

FIGURE 8.4   Information in the user interface structured according to a functional style.

performed information (e.g. department store item) search tasks using the assigned user interface. Participants were instructed to perform the information search tasks as quickly as possible without sacrificing accuracy. There were three consecutive trials for each subject in which 30 items were listed. Each subject was given the three trials in the same sequence. The tasks to be performed were exactly the same in each session but with items listed in different orders. American and Taiwanese subjects required approximately 60 minutes to perform the tasks. Mainland Chinese required approximately 90 minutes. Performance time and errors were recorded. The subject committed an error when he or she navigated to a category in the store that did not contain the product he or she was asked to search for.

On the completion of the information search tasks, each participant was given a memory free recall test in order to assess information retrieval. Participants were not told about the memory test before or during the tasks to ensure that no participant

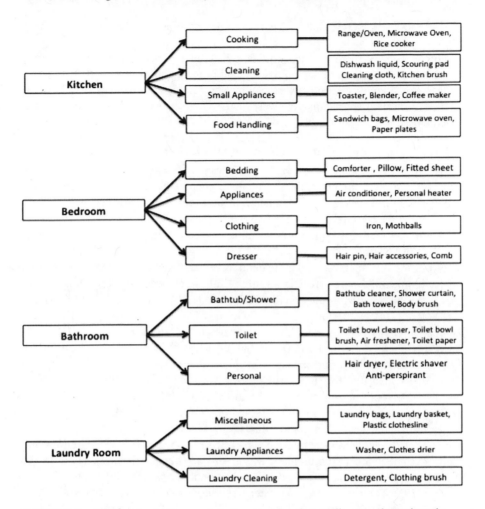

FIGURE 8.5   Information in the user interface structured according to a thematic style.

would try to memorize information during task performance. The memory test required participants to freely recall and write down as many of the categories, subcategories, and items as possible in 10 minutes time. The number of items correctly retrieved from memory was recorded.

8.4.4   RESULTS

8.4.4.1   Knowledge Representation (Abstract vs. Concrete)

For Mainland Chinese subjects in the first experiment (Choong and Salvendy, 1999) there was a small, but significant difference in performance time on Trial 3 (F (1,36) = 8.81, p = 0.0054) using the abstract representation. For the third trial, the mean performance time of Mainland Chinese subjects with abstract assignment (Mean = 200.765, SD = 54.7) was 9% slower than that of subjects with concrete assignment

(Mean = 181.926, SD = 36.098). There were no differences in errors for Mainland Chinese subjects.

For Americans, there was no significant difference in performance time or errors between the abstract and concrete knowledge representation conditions. Chinese in Taiwan studied in the second experiment by Rau, Choong, and Salvendy (2004), performed much like Americans, relatively unaffected by the knowledge representation scheme used in the user interface.

8.4.4.2   Information Structure (Functional vs. Thematic)

In the first experiment, conducted by Choong and Salvendy (1999) and using American and Mainland Chinese subjects, there was a significant interaction between Cultural Group and Information Structure in Trial 1 for performance time (F $(1,76)$ = 4.43, p = 0.0387) and for errors (F $(1,76)$ = 23.35, p = 0.0001). There were no significant Cultural Group × Information Structure interactions for the second and the third trials.

Exploring this interaction in more detail revealed that Mainland Chinese subjects made significantly more errors in all three trials when searching for items with the user interface organized functionally. (Trial 1: F $(1,36)$ = 18.38, p = 0.0001; Trial 2: F $(1,36)$ = 29.00, p = 0.0001; Trial 3: F $(1,36)$ = 24.58, p = 0.0001). In Trial 1, the number of errors committed with the thematic user interface (Mean = 23.5, SD = 7.58) was 39.80% less than with functional UI (Mean = 39.0, SD = 13.85). Similarly, in Trial 2, the number of errors using the thematic UI (Mean = 12.0, SD = 6.19) was 51.52% less than that for functional UI (Mean = 24.8, SD = 8.96). In Trial 3, the number of errors for subjects using the thematic UI (Mean = 6.1, SD = 5.45) was 58.70% smaller than that for the functional one (Mean = 14.7, SD = 6.18). On Trial 2, performance time for Mainland Chinese subjects was significantly faster with the thematically organized user interface (F $(1,36)$ = 4.45, p = 0.0419). The Mainland Chinese subjects using the thematic user interface (Mean = 231.707, SD = 44.965) were 15.60% faster than those using the functional user interface (Mean = 274.527, SD = 77.183). Their time also was faster with the thematic user interface on Trials 1 and 3, but not significantly so.

American subjects had no significant differences in mean performance times between the functional and thematic user interfaces. However, on Trial 1, these subjects had significantly fewer errors when using the functional compared to the thematic UI (F $(1,36)$ = 8.62, p = 0.0058). The number of errors committed in the first trial by the American subjects with the functional UI (Mean = 22.1, SD = 8.51) was 64.25% smaller than that committed by those subjects with thematic UI (Mean = 36.3, SD = 20.44). In Trials 2 and 3, they showed no significant difference in errors.

The second experiment, conducted by Rau, Choong, and Salvendy (2004) applied this identical paradigm and UI design to Chinese subjects in Taiwan. They found no significant effect of information structure on the searching time of these Taiwan Chinese. Looking at the effects on errors, they found no significant difference in the first trial. However, these subjects made significantly fewer errors using the thematic user interface in Trial 2 (F $(1,36)$ = 3.27; p = 0.0789) and Trial 3 (F $(1,36)$ = 5.04; p = 0.0310). In Trial 2 the Taiwan Chinese subjects using the thematic UI made 39.9% less errors (Mean=13.7, SD=11.27) than the participants who used the functional

UI (Mean=22.8, SD=18.13). In Trial 3, subjects using the thematic UI made 61.5% fewer errors (Mean=6.5, SD=5.45) than subjects with the functional UI (Mean=16.9, SD=16.13).

Finally, on the item recall task performed by all subjects after the experiment, only Mainland Chinese were significantly affected by the information structure they had used during the experimental trials (F (1,36) = 5.23, p = 0.0282). Subjects who had used the thematic UI during the experiment recalled slightly more items 12.58%, (Mean = 34.9, SD = 5.28) than those who had used the functional assignment (Mean = 31.0, SD = 5.80). Taiwan Chinese showed no differences in recall.

8.5   CONCLUSIONS

From these two experiments, we can conclude, first, that Chinese from both the Mainland and Taiwan will perform better using a thematic as opposed to a functional information structure. This conclusion is particularly strong for Mainland Chinese who had significantly more errors on all three trials when they were forced to interact with a user interface designed with a functional, rather than their preferred thematic information structure. For Chinese users in Mainland China and Taiwan, a user interface built on a thematic information structure most likely will be advantageous in a range of product, service, and system designs.

Also, we can conclude that, for novice users or people using an application for the first time, it is important to provide a user interface with a culturally compatible information structure. Americans were uncomfortable with a thematically-oriented user interface on their initial exposure to it, as evidenced by more errors on the first trial in the experiment. They eventually adapted to it on subsequent trials. However, what if this was a real e-commerce website and the subject's initial experience with it determined his or her willingness to return again and purchase a product? If the information structure seems odd and the user makes many erroneous attempts to find the item they seek, they will be much less inclined to return and try it again. There may not be a second or third trial in the real world of e-commerce. The user is lost as a customer of that e-commerce company.

Finally, there were few clear findings to support the notion that Chinese will prefer a concrete representation of knowledge in the UI over an abstract one. Perhaps the graphical metaphor of the department store used in these experiments was not sufficient to effectively support their cognitive style and natural knowledge representation preferences. However, the theory remains compelling. Thus, this is a good subject for further research.

8.6   APPLICATION

Table 8.1 shows some suggested design strategies for knowledge representation and information structure for users in the United States, Taiwan, and Mainland China.

TABLE 8.1

Advantageous System for Different Cultures

		Chinese	
Components	American	Mainland China	Taiwan
Knowledge representation	Abstract or concrete	Concrete	Abstract/concrete
Interface structure	Functional or thematic	Thematic	Thematic

REFERENCES

Anderson, R., Anderson, R., and Deibel, K., 2004. Analyzing concept groupings of introductory computer programming students. Technical Report. University of Washington, November.

Chiu, L.H. 1972. A cross-cultural comparison of cognitive styles in Chinese and American children. *International Journal of Psychology* 7: 235–242.

Choong, Y.Y. 1996. Design of computer interfaces for the Chinese population. Doctoral dissertation, Purdue University, August.

Choong, Y.Y., and Salvendy, G. 1999. Implications for design of computer interfaces for Chinese users in Mainland China. *International Journal of Human-Computer Interaction*, 11: 29–46.

Cyr, D., and Trevor-Smith, H. 2004. Localization of Web design: An empirical comparison of German, Japanese, and United States Web site characteristics. *Journal of the American Society for Information Science and Technology*, 55(13): 1199–1208.

Fang, X., and Rau, P.L.P. 2003. Culture differences in design of portal sites. *Ergonomics*, 46(1–3): 242–254.

Hall, E.T. 1984. *Dance of Life: The Other Dimension of Time*. Yarmouth, ME: Intercultural Press.

Hofstede, G. 1980. *Culture's Consequences: International Differences in Work-Related Values*. London: Sage.

Kim, J.H., and Lee, K.-P. 2007. Culturally adapted mobile phone interface design: Correlation between categorization style and menu structure. *Mobile HCI 2007*: 379–382.

Kim, J.H., Lee, K.P., and You, I.K. 2007. Correlation between cognitive style and structure and flow in mobile phone interface: Comparing performance and preference of Korean and Dutch users. In *Usability and Internationalization*, ed. N. Aykin, Part 1, HCII 2007, LNCS 4559, 531–540. Heidelberg: Springer-Verlag.

Kralisch, A., Eisend, M., and Berendt, B. 2005. Impact of culture of Website navigation behaviour. In *Proceedings of the 11th International Conference on Human-Computer Interaction,* Las Vegas, July 22–27. Mahwah, NJ: Lawrence Erlbaum.

Lin, Y.T. 1939. *My Country and My People*. New York: John Day.

Marcus, A., and Gould, E.M. 2000. Crosscurrents: Cultural dimensions and global Web user-interface design. *Interactions*, 7: 32–46, ACM Publisher, www.acm.org, July/August.

Morkes, J., and Nielsen, J. 1997. Concise, scannable and objective: How to write for the web. http://www.useit.com/papers/webwriting/writing.html.

Nawaz, R., Plocher, T., Clemmensen, T., Qu, W., and Sun, X. 2007. Cultural differences in the structure of categories in Denmark and China. Working paper 03-2007, Department of Informatics, Copenhagen Business School.

Nielsen, J. 1997. Changes in Web usability engineering since 1994. http://www.useit.com/alertbox/9712a.html.

Nisbett, R.E. 2003. *The Geography of Thought: How Asians and Westerners Think Differently and Why*. New York: Free Press.

Northrop, F.S.C. 1946. *The Meeting of East and West*. New York: Macmillan.

Plocher, T.A., Garg, C., and Krishnan, K. 1999. Cross-cultural issues in the business aviation cockpit. Technical report prepared for Honeywell Business and Commuter Aviation Systems, June.

Plocher, T.A., Zhao, C., Liang, S.M., Sun, X., and Zhang, K. 2001. Understanding the Chinese User: Attitudes Toward Automation, Work, and Life. In *Proceedings of the Ninth International Conference on Human-Computer Interaction*, New Orleans, LA, August.

Rau, P.-L.P., Choong, Y.Y., and Salvendy, G. 2004. A cross cultural study on knowledge representation and structure in human computer interfaces. *International Journal of Industrial Ergonomics*, 34(2): 117–129.

Rau, P.-L.P., and Liang, S.-F.M. 2003a. A study of cultural effects on designing user interface for a Web-based service. *International Journal of Services Technology and Management*, 4(4–6): 480–493.

Rau, P.-L.P., and Liang, S.-F.M. 2003b. Internationalization and localization: Evaluating and testing a website for Asian users. *Ergonomics*, 46: 255–270.

Stewart, E.C., Bennett, M.J. 1991. *American Cultural Patterns: A Cross-Cultural Perspective*. Yarmouth, ME: Intercultural Press.

Weinschenk, S., Jamar, P., and Yeo, S.C. 1997. *GUI Design Essentials*. New York: Wiley.

Yang, K.S. 1986. Chinese personality and its change. In: Bond, M.H. (Ed.), *The Psychology of the Chinese People*. New York: Oxford University Press, pp.106–170.

Yang, K.S., Tsai, S.G., Hwang, M.L. 1963. Rorchach responses of normal Chinese adults: III. Number of responses and number of refusals. *Psychological Testing (Taipei)* 10: 127–136.

Zhao, C. 2002. Effect of information structure on performance of information acquiring: A study exploring different time behavior: Monochronicity/polychronicity. PhD dissertation, Institute of Psychology, Chinese Academy of Sciences, May.

9 Physical Ergonomics and Anthropometry

9.1 INTRODUCTION TO THE PROBLEM

9.1.1 ANTHROPOMETRY

The study of body sizes and other associated characteristics is generally referred to as anthropometry (Lehto and Buck, 2008). Anthropometry involves the measurements of body size, shape, strength, mobility and flexibility, and working capacity (Pheasant and Haslegrave, 2006).

Anthropometric measurements are essential when designing devices, equipment, and systems for users. Humans vary in body dimensions, shapes, and other characteristics, thus ergonomic design requires an understanding of the variability of human beings. Depending on the database consulted, the measurements included may range from simply stature and weight, to thousands of measurements, including multiple postures and detailed facial measurements.

Anthropometric data can be classified by the sample of the subjects: military soldiers or civilians. The anthropometric information about the military soldiers has a long history and is rather complete (Kroemer, Kroemer, and Kroemer-Elbert, 1990). The military has always had a particular interest in the body dimensions of soldiers in order to provide proper fitting of uniforms, armor, and other equipment. The recently published Human Integration Design Handbook (HIDH), NASA/SP-2010-3407 (NASA, 2010) provides a good overview of the anthropometry as well as provides guidance for human factors design especially for human space flight programs and projects. However, military data should be used with caution when applied to a civilian population because of the selection biases of the young and healthy sample. In the past ten years, many institutes, libraries, and commercial companies across the world conducted large surveys of anthropometric data of the civilians in different nationalities. The anthropometric data of civilians from different nationalities can be used for designing products for people from different nationalities.

9.1.2 MOVEMENT AND REACH ZONE

Classical anthropometric data provide information on static or structural dimensions of the human body in standard postures. However, these data cannot describe functional performance capabilities, such as reach capabilities and movements. When performing a task, humans do not maintain standard and static postures. Furthermore, human movement varies from whole-body movement (e.g., locomotion

TABLE 9.1

Human Body Movement Databases

Number	Author	Country	Data
1	Barter, Emmanuel, and Truett, 1957	United States	19 joint movements
2	Lehto and Buck, 2008	United States	21 joint movements
3	NASA, 2010	United States	25 joint movements
4	Hu et al., 2006	China	18 joint movements
5	DINED	Netherlands	11 joint movements

or translation) to partial-body movement (e.g., controlling a joystick with the right arm), to a specific joint or segment movement (e.g., pushing a button with a finger while holding the arm steady) (NASA, 2010). Thus, the static postures cannot provide the advantages of dynamic posture that are involved in the design.

An ergonomic designer must be familiar with how the human body moves, especially when designing workspaces (Lehto and Buck, 2008). In ergonomic design, the movements often of interest are the movements around a joint, for instance shoulder movement, wrist movement, hip movement, and ankle movement. Table 9.1 provides a summary of the resources of human body movement data across the world.

Human body movement data can help designers to determine the proper placement and allowable movement of controls, tools, and equipment (NASA, 2010). Body movement data can be combined with static body dimensions to calculate the movement ranges and reach zones in the workplace.

In ergonomic design, two aspects of reach should be carefully designed: zones of convenient reach (ZCRs), and the normal working area. The zone of convenient reach is the appropriate zone or space in which an object may be reached conveniently by an individual. The normal working area is described as a comfortable sweeping movement of the upper limb, about the shoulder, with the elbow flexed to 90 degrees or a little less (Pheasant and Haslegrave, 2006). The data of ZCR for a full grip and the coordinates of the normal working area can be found in the book by Pheasant and Haslegrave (2006). NASA (2010) provides the data for grasp reach limits with right hands for Americans.

9.2 GUIDELINES FOR APPLICATION OF ANTHROPOMETRY TO CROSS-CULTURAL DESIGNS

This section describes the general considerations for database selection, identifying the user population, and the methods of applying anthropometric data into designs.

Ergonomics and anthropometry are very important in the creation of usable products. Anthropometric data drives the guidelines for the design of a product (NASA, 2010). First, knowing the target user population determines which database to use. Second, once the target population is defined, designers must decide on the range of the personnel in that population who will be operating and maintaining the product.

9.2.1 Select Appropriate Database

9.2.1.1 Why?

Currently, there are many sources to find anthropometric data, for example in papers in scientific journals, in compilations such as the Anthropometric Source Book (Webb Associates, 1978), ADULTDATA (and its companion volumes CHILDATA and OLDERADULTDATA [Norris and Wilson, 1995; Peebles and Norris, 1998; Smith et al., 2000]), and in the anthropometric database such as CAESAR®, PeopleSize 2000. Table 9.2 introduces the most used online anthropometric databases for different nations.

9.2.1.2 How?

For years, the measurement of body dimensions used traditional tools to generate one-dimensional measurement data. The tools included calipers, measuring tapes, anthropometer, weight scale, sliding compass, head spanner, and other similar instruments. In recent decades, the emerging and fast development of three-dimensional (3D) scanning technology greatly changed the way of anthropometric studies. The three-dimensional body scanning technology has many advantages over the traditional measurement system. It is capable of capturing hundreds of thousands of points in a few seconds. Moreover, it provides details about the surface shape and three-dimensional locations of measurements relative to each other. The digitized measurements can be easily transferred to computer-aided design and manufacturing tools. In all, the noncontact, instant, and accurate three-dimensional measurement has made anthropometric studies more convenient to conduct.

In recent years, several national-level three-dimensional anthropometric surveys have been conducted. They provide online databases that either can be freely used or have to be purchased for use. The Civilian American and European Surface Anthropometry Resource (CAESAR®) collected anthropometric information from 2,400 U.S. and Canadian and 2,000 European civilians ages 18 to 65. They provide a three-dimensional as well as one-dimensional database on 40 anthropometric measurements.

The UK National Sizing Survey (Size UK) collected data from 11,000 subjects ages 16 to 90-plus years from UK populations. It used 3D whole-body scanners to automatically extract 130 body measurements for standing and seated poses. Size USA conducted a comprehensive sizing survey of the U.S. population by 3D body scanning technology. It collected data of nearly 11,000 subjects from 12 locations across the United States. Size China collected the head and face sizes of Chinese population ages 18 to 70 in six different locations in mainland China. It is the first 3D database of Chinese head and face sizes which can be used for international manufacturers and designers. In Japan, the Research Institute of Human Engineering for Quality Life conducted many 3D anthropometric measurements on the Japanese population.

Different from the database above, the World Engineering Anthropometric Resource (WEAR) is an international organization that organizes a group of interested experts involved in the application of anthropometry data for design purposes.

TABLE 9.2

Summary of Anthropometric Database for Different Nations

Database	Institute/Author	Population	Age	Data	Methodology	Link
WEAR	Global partners	Depends on different studies	Depends on different studies	Depends on different studies	WEAR is a group of interested experts involved in the application of anthropometry data for design purposes.	http://wear.io.tudelft.nl/
PeopleSize 2000	Open Ergonomics Ltd.	American, Australian, Belgian, British, Chinese, French, German, Japanese, Swedish	Infant, child, and adult	Body measures	289 individual body measurement dimensions	http://www.openerg.com/psz/index.html
DINED	Delft University of Technology	Dutch; International: North American; Latin American (Indians); Latin American (rest); North Europe; Central Europe; Eastern Europe; South East Europe; France; Spain and Portugal; North Africa; West Africa; South East Africa; Middle East; North India; South India; South China; South East Asia; Australia (European); Japan	2 to 80+ Dutch civilians	Body measures, force exercise, and joint excursion	Long-term project collecting data from different nations. Latest version for Dutch adults is DINED 2004, for Dutch elderly is GERON 1998 and for Dutch children is KIMA 1993. The version for international is 1989.	www.dined.nl

Name	Organization	Population	Sample	Measurements	Description	URL
AnthroKids	Information Technology Laboratory (ITL) at the National Institute of Standards and Technology (NIST) and the Consumer Product Safety Commission (CPSC)	North American	Children	Body measurements	Two studies performed in the years 1975 and 1977	http://ovrt.nist.gov/projects/anthrokids/
CAESAR®	Government and industry	North American; European	18 to 65, civilian	40 body dimensions	From April 1998 to early 2000, the project collected data on 2,400 U.S. and Canadian and 2,000 European civilians.	http://store.sae.org/caesar/
DINBelg 2005	Several Belgium schools	Belgian	2 to 80, civilian	Body measurements	The aim of DINBelg 2005 is to gather up-to-date anthropometric dimensions of the Belgian population.	http://www.dinbelg.be/anthropometry.htm
Size China	Hong Kong Polytechnic University	Chinese	18 to 70, civilian	Head and face size	From 2006 to 2007, the project collected data on 2,000 male and female volunteers in six diverse locations in China.	http://www.sizechina.com
Size Japan	Research Institute of Human Engineering for Quality Life	Japanese	Civilian	Body measures	—	http://www.hql.jp/

They collected different anthropometric data across the world as well as the methods in a wide variety of innovative applications. Its aim is to develop data models and software tools of an online worldwide information system for utilizing the latest anthropometric databases in engineering environments.

Besides databases listed above that can provide up-to-date 3D measurements, there are many anthropometric databases that provide one-dimensional measurements for researchers and designers. For example, PeopleSize 2000 provides 289 body measurement dimensions of American, Australian, Belgian, British, Chinese, French, German, Japanese, and Swedish populations. DINED provides an extensive database for the Dutch population ages 2 to 80-plus years. The DINBelg 2005 provides body measurements for the Belgian population. The AnthroKids provides anthropometric data of children in North America.

Due to the expensive and time-consuming surveys necessary to collect representative and comprehensive anthropometric data, the data are usually not up-to-date and may not best represent the current user population. Secular changes occur quite rapidly in populations; thus, predictions for future secular trends and corrections may need to be made when using the anthropometric data from databases.

In summary, best practice for choosing a database for cross-cultural product developments requires the following:

- Choose a database that is:
  - Closely representative of the user population
  - Current
  - Large enough to overcome statistical issues
- Consider the user's nationality, age, gender, physical condition, and education level
- Consider the historical changes that have occurred in anthropometric dimensions from generation to generation
- If possible, conduct user surveys that are sufficiently large (at least 1000 subjects) to account for population variances

9.2.2 CAREFULLY CHOOSE THE USER POPULATION

9.2.2.1 Why?

Choosing the user population is a very important step in the design process. It is the basis of the foregoing discussion and analysis. Users' age, gender, ethnicity, and other special characteristics should be considered. Apart from a few special small populations which are completely known, it is rare to have anthropometric data directly applicable to a target population (Pheasant and Haslegrave, 2006). For example, it is difficult to have anthropometric data for production line operators within a particular factory. Thus, the general (civilian) population anthropometric statistics for the appropriate nationality are normally used.

People from different nationalities vary in their body dimensions. Thus designers should carefully consider the national differences in body dimensions for people in different nationalities. Lin, Wang, and Wang (2004) compared the anthropometric characteristics among four populations in East Asia. They found significant morphological differences among these peoples on the following four aspects. First, the Mainland Chinese body shape has a narrower body with mid-range limbs. Second, the Japanese body shape is wider with shorter limbs. Third, the Korean body shape is mid-range among the four peoples, but the upper limbs are longer. Fourth, the Taiwanese body shape has wide shoulders and narrow hip with large hands and long legs.

9.2.2.2   How?

In summary, best practice is design in a way that adapts to regional variations in anthropometry. For countries with very large territories, anthropometric data related to physical stature may vary considerably from area to area. Take China as an example. Stature, chest breadth, and weight all vary between the six major areas of China: the Northeast, the North, Northwest, Southwest, Southeast, and Central (GB 10000-88 China). Also, regional differences in climate and environment will affect what is required in terms of the comfort and fit, as well as thermal properties of clothing items such as firefighter protective clothing.

9.3   CASE STUDY 1: RAPID PRELIMINARY HELMET SHELL DESIGN BASED ON THREE-DIMENSIONAL ANTHROPOMETRIC HEAD DATA

We now introduce a study conducted by Liu, Li, and Zheng (2008). They developed a rapid preliminary helmet shell design method based on 3D anthropometric head data.

9.3.1   BACKGROUND

Helmets are personal protective equipment widely used in construction, manufacturing, sports, riding, and so forth. For some occupations, such as construction, workers wear helmets for a long period of time, which causes discomfort and pain. Thus, integration of lightweight design and comfort is important.

The traditional development process for the helmet shell has two major disadvantages. The first is that the design does not always fit the human head because the traditional anthropometric information available to most designers is misleading and can lead to poor helmet sizing. The second disadvantage is that all the steps in the traditional process are expensive and time consuming, and a design is difficult to modify. Thus, the traditional method cannot meet the requirements of individual helmet design.

Computer-aided design (CAD) has shown potential in helmet design. Computer-generated models of 3D head data could only be used to visually analyze whether or not the helmet is being designed to fit the head. If the head model needs to be changed, it will involve a tedious and time-consuming task of helmet modification.

9.3.2   OBJECTIVE

The object was to rapidly develop a preliminary helmet shell design.

9.3.3 METHOD

First, a 3D anthropometric head scan was taken as the design reference to obtain satisfactory shape fitting. A 3D head model was generated from the 3D head scan of the intended user or representative user of an intended population group.

Then, a semiparametric surface modeling method based on a 3D head model was proposed for the rapid generation of helmet shells. The method was based on simply inputting several parameters related to helmet protection, size, and shape requirements and adjusting several key curves.

After that, to facilitate application of the semiparametric model, a computer-aided modeling tool, Helmet Design, was developed to quickly generate the helmet shell.

At last, the new design by the proposed method was compared with the existing design by the traditional method with regard to weight, centroid, and moments of inertia to test the effectiveness of the proposed method.

9.3.4 RESULTS

A redesign of an existing safety helmet with the proposed computer-aided helmet design tool was compared with the traditional method. The aim was to demonstrate the improvement of the proposed method on the helmet's comfort, especially with regard to weight reduction.

The geometric and physical properties related to comfort were evaluated between the original and the redesigned shell. The results showed that the redesigned helmet shell was visibly smaller in height than the original shell. It weighed 12.8% less than the original. When using the traditional method, it was hard to control the protection distance accurately without the reference of a 3D head model. Thus, the designer was inclined to design a conservative space between the shell and the head, which may explain the reason the original shell had an excessive height. The extra space in the upper-front and upper-back areas of the original shell also resulted in a heavier helmet.

In addition, the centroid position relative to the 3D head model and the moments of inertia regarding the centroid of the helmet shell were compared. The results showed that the centroid of the redesigned helmet shell was 10.9 mm lower than the original (z-axis direction). The moments of inertia of the helmet shell were also decreased a lot. These results suggested improvement of stability for the redesigned helmet.

9.3.5 CONCLUSIONS

The results showed that a preliminary design of a safety helmet shell in reference to 3D anthropometric measurement of a human head can be generated rapidly by using the computer-aided tool based on the proposed semiparametric helmet shell model. In addition, the proposed method improved and speeded up the development process of customized safety helmets. The integrated helmet-head model also provided an easier evaluation of a certain design with ergonomics considerations. The redesign of an existing safety helmet also demonstrated the effectiveness of the proposed method.

In conclusion, the proposed method greatly supported the customized design of safety helmets such as that required to accommodate various regional populations. It improved wearing comfort and provided better protection for the users.

9.4 CASE STUDY 2: ANTHROPOMETRIC MEASUREMENT AND CHINESE MEDICAL ACUPUNCTURE

Anthropometric measurement is an indispensable item in human life. The anthropometric data are used for manufactured product and garment design.

The traditional anthropometric measurement method requires manual measurements by domain experts, combined with precise measurement devices. Although the traditional method is easy and convenient to use, it still relies on manual operations that are inefficient and prone to errors. Another measurement method using 2D image-based measurements adopts two or more photographic images to do the measurements. Usually, two images are captured: front and side. The weakness of 2D image-based measurement is the complicated presetting, camera setting, and calibration, and too much manual operation. This makes this method not easy for people to use. A more advanced measurement using 3D laser scanning technique can provide a standard model of a digitized human body shape and opens up opportunities to extract new measurement for quantifying the body shape. However, the 3D scanning equipment is not always available, and the collected data still need complicated analysis to calculate anthropometric data.

With the fast development of the Internet, the traditional method, 2D image-based method, as well as the 3D scanning method may not very well fit the quick change in the consumption pattern, for example, the virtual shop online, where customers can make a purchase anytime and anywhere via the Internet. Thus, a method for anthropometric measurement without a complex setting, no high-price measurement instruments, and no experience required is highly needed.

In the following case study, we will describe a research conducted by Lin, Chien, and Chiu (2010). They presented a simplified 2D image-based anthropometric measurement approach by using the Chinese medicine acupuncture theory and human body slice model, and they illustrated that the new approach was quite comparable to traditional measurement performed by skilled anthropometrists. Before introducing Lin, Chien and Chiu's work, we will first introduce the bone length measurement and Chinese medicine acupuncture, and discuss the relationship between Chinese medicine acupuncture and anthropometric measurement by two related studies: Li et al. (2008) and Wu (2011).

9.4.1 BACKGROUND

Traced back to ancient times, Chinese people used the "bone length measurement" to get the size and length of the human body. Bone length originated from "Lingshu Bone Length," which is an ancient medical book in the Warring States Period. In ancient China, bone length was widely used as the basis for positioning of acupuncture point selection. Therefore, using bone length measurement for the measurement of acupoints location is called the bone length method. Figures 9.1 and 9.2 illustrate

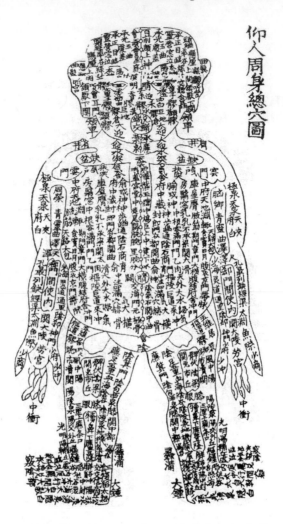

FIGURE 9.1 The position of acupuncture points on the human body (face up).

the position of acupuncture points on the human body (China Association of Chinese Medicine, 2007). A more detailed description of bone length measurement is as follows. The ancients compared the body size and further transferred the size by standard units after a lot of observation, measurement, and anatomical work on the human body. For instance, the standard human height is set as 75 units, and then the length and width between two bones can be converted into certain units. One unit equals 1 inch, 10 units equal 1 foot. The way to determine different human body sizes according to proportion is the bone length measurement. The ancients measured the bone length after the actual measurement. With the fast development of measurement techniques today, a question is raised: can the proportion revealed by bone length be verified by today's anthropometric measurements?

Li et al. (2008) conducted a study focused on the location of head acupoints by bone length measurement. They compared the differences between bone length

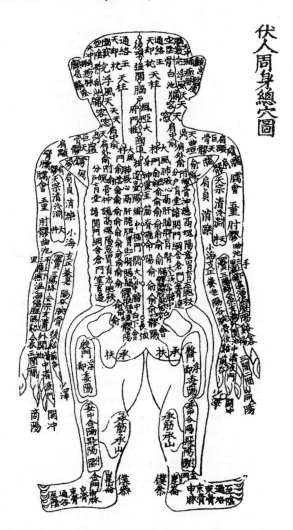

FIGURE 9.2 The position of acupuncture points on the human body (face downward).

measurements on the human head and probe proportion relation of the position of head points. Their study measured several bone lengths of 100 healthy adults ages 18 to 25 years using standard measurement instruments. The measurements included height, body weight, distance from Shenting (GV 24) to Toulinqi (GB 15), from Shenting to Touwei (ST 8), between bilateral Touwei (ST 8), between the two Mastoid, from Yintang (EX-HN 3) to front hairline, from hairline to Naohu (GV 7), Naohu to Fengfu (GV 16), and Fengfu to the middle or rear hairline.

The results revealed that some acupuncture points were consistent with the bone length measurement. For instance, Toulinqi can be located at the middle point of the connecting line of Shenting and Touwei. Their results also indicated that some acupuncture points were inconsistent with the bone length measurement. For instance, bilateral Touwei distance was not the same as the bone length measurement between two Mastoid. More recently, Wu (2011) reviewed and compared the relationship of anthropometric measurement and bone length measurement. He also concluded that many anthropometric measurements used today were consistent with the acupuncture points in bone length measurement. By using anthropometric method, we cannot only verify the proportions of the human body discovered by ancient people but also adjust certain inconsistent bone lengths.

9.4.2 OBJECTIVE

The objective was to develop an approach which has fewer constraints, more simplified operations, and as comparable as traditional measurement (e.g., accuracy, precision, and repeatability).

9.4.3 METHOD

The new approach proposed by Lin, Chien and Chiu (2010) is as follows. First, they used the Chinese medicine acupuncture theory to locate the critical position and replace the manual marking or other feature extraction methods. Five mostly used body measurement data in manufacturing and general application were selected and used: shoulder length, chest circumference, waist circumference, hip circumference, and leg length. These five body measurement data can be located by six acupuncture points: Lian-Quan, Tian-Tu, Chien-Yu, Shan-Zhong, Shen-Que, and Qu-Qu. For example, the position of the Qu-Qu acupuncture point is near the upper edge of the pubic symphysis and could be defined as the position of hip. After locating the critical position, the 2D anthropometric data can be obtained by direct measurement, which is calculated by the distance between two acupuncture points. Next, for circumferences measurement, the human body slice model was used to approximate the circumference shape, and the piecewise Bezier curve was used to approximate the circumference curve. At last, they used a compensation system of garment thickness to amend the measurement data by directly measuring subject wearing clothes. By using the new approach, subjects were not required to wear as little as possible for measurement.

9.4.4 RESULTS

The proposed approach was compared with 2D image-based measurement using the manual measurement (or called traditional measurement). Two experiments were conducted. The first experiment was to calculate the accuracy of the measurement system with less influence of ferment. Participants in the first experiment wore short-sleeve T-shirts. The second experiment tested the accuracy of measurement with manifold garments. Participants in the second experiment wore four kinds of garments: short T-shirt, long T-shirt, thin jacket, and thick jacket. Each participant in the two experiments was measured once by each method. The accuracy, precision, and reliability were evaluated. The results indicated the new approach was comparable to the traditional measurement method performed by skilled anthropometrists in terms of accuracy, precision, and repeatability.

9.5 CONCLUSIONS

The new approach to anthropometric measurement based on 2D imaging was easy to take using a digital camera without complicated settings, and can attain high accuracy and reliability compared to manual measurement. The new approach can be used in large-scale anthropometric investigation of population and merchandising (e.g., customized clothing made by using photographs only and without the customer's physical presence).

REFERENCES

Barter, T., Emmanuel, I., and Truett, B. 1957. *A Statistical Evaluation of Joint Range Data*. Ohio: Wright Patterson Air Force Base.

China Association of Chinese Medicine. 2007. *Traditional Chinese Medicine 100 Masterpieces: Acupuncture and Moxibustion Volume*, 41–42. Beijing: Huaxia. (Reprinted from J. Yang, *Compendium of Acupuncture and Moxibustion*, Vol.1, 1601.)

Hu, H., Li, Z.Z., Yan, J., Wang, X., Xiao, H., Duan, J., et al. 2006. Measurements of voluntary joint range of motion of the Chinese elderly living in Beijing area by a photographic method. *International Journal of Industrial Ergonomics*, 36: 861–867.

Kroemer, K.H.E., Kroemer, H.J., and Kroemer-Elbert, K.E. 1990. *Engineering Physiology: Bases of Human Factors/Ergonomics*, 2nd edition. New York: Van Nostrand Reinhold.

Lehto, M.R., and Buck, J.R. 2008. *Introduction to Human Factors and Ergonomics for Engineers*. New York: Taylor and Francis.

Li, M., Hu, L., Cai, R., Chen, W., Meng, Y., Wu, Z., et al. 2008. A report on location of head acu-points by bone-length measurement in 100 persons. *Chinese Acupuncture*, 28: 273–275.

Lin, S., Chien, S., and Chiu, K. 2010. The 2D image-based anthropologic measurement by using Chinese medical acupuncture and human body slice model. *International Journal of Computer Science and Information Security*, 8: 20–29.

Lin, Y., Wang, M.J., and Wang, E.M. 2004. The comparisons of anthropometric characteristics among four peoples in East Asia. *Applied Ergonomics*, 35: 173–178.

Liu, H., Li, Z.Z., and Zheng, L. 2008. Rapid preliminary helmet shell design based on 3D anthropometric head data. *Journal of Engineering Design*, 19: 45–54.

NASA. 2010. *Human Integration Design Handbook*. Washington, DC: National Aeronautics and Space Administration.

Norris, B., and Wilson, J.R. 1995. Childata. *The Handbook of Child Measurements and Capabilities: Data for Design Safety*. London: Department of Trade and Industry.

Peebles, L., and Norris, B. 1998. ADULTDATA. *The Handbook of Adult Anthropometric and Strength Measurements: Data for Design Safety*. London: Department of Trade and Industry.

Pheasant, S., and Haslegrave, C.M. 2006. *Bodyspace: Anthropometry, Ergonomics, and the Design of Work*, 3rd edition. Boca Raton, FL: Taylor and Francis.

Smith, S., Norris, B., and Peebles, L. 2000. OLDER ADULTDATA. *The Handbook of Measurements and Capabilities of the Older Adult: Data for Design Safety*. London: Department of Trade and Industry.

Webb Associates. 1978. *Anthropometric Source Book*. NASA Reference Publication No. 1024. Lyndon B. Johnson Space Center, Houston, TX: U.S. National Aeronautics and Space Administration.

Wu, X. 2011. Bone length measurement and anthropometric measurement. *Chinese Journal of Basic Medicine in Traditional Chinese Medicine*, 17: 111–112.

Yang, J. 1601. *Compendium of Acupuncture and Moxibustion* (Vol. 1, pp.148). China Association of Chinese Medicine, 2007.

Section III

Methodology

10 Weaving Culture into the Product Development Process

10.1 INTRODUCTION

In Section I, we reviewed the theoretical rationale for carefully considering culture in user interface design. Section II presented guidelines to follow when designing a user interface for cultures nonnative to yours. But most information technology (IT) applications are developed in corporations, and corporations have standardized processes for product development. This begs the practical question of how one incorporates cross-cultural user interface design guidelines and methods into a standard product development process.

This chapter provides an overview of cross-cultural user interface design and the product development process. Figure 10.1 shows a general engineering process for product development. It also shows where in the product development process key cross-cultural studies should be conducted and documented in order to develop a successful product. Each of these process steps and supporting cross-cultural studies will be discussed in this chapter. More detailed guidance for conducting critical cross-cultural design studies is provided in Chapter 11 (User Needs Research: Understanding the Lay of the Land), Chapter 12 (Gaining User Acceptance in Specific Cultures), Chapter 13 (International Usability Evaluation), and Chapter 14 (Heuristic and Guildelines-Based Evaluation in Cross-Cultural Design).

10.2 PHASE 1: IDENTIFY CUSTOMER (USER) REQUIREMENTS

Typically, Phase 1 of the product development process is focused on the questions of "what should we build?" and "to whom should we sell it?" In most corporations, this initial phase of development is led by the marketing department, with contributions from the technology or engineering teams in the area of technology assessment. The objective is to better understand some particular customer need or problem and broadly conceive of a potential solution for which the technology is feasible. Then they assess the size of the potential market for the solution and whether or not it is large enough to be worth the always significant investment in product development. Phase 1 sets the stage for the rest of the product development, in fact, determines whether or not product development will be undertaken at all.

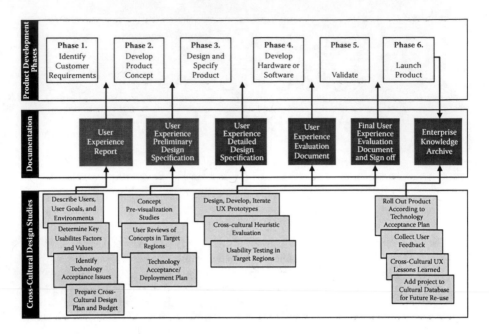

FIGURE 10.1 Cross-cultural design studies and documentation overlaid on a general engineering process for product development.

Traditionally, this initial stage of product development was viewed as the purview of the marketing department. Often, that leads to a focus on "the customer," not the end user. Customers are critically important to understand and satisfy. They set the "big picture" for a problem and solution and are the ones who eventually pay for the product. But, usually, they are not the ones who will actually use the product. For example, if I am developing a new home security system, my customer likely will be a distributor or a security company, not the homeowner. If I am developing a new alarm display for an oil refinery, the IT manager of the refinery is the likely customer and the one who will "pay" for it. However, the users are the plant engineers and control room operators. All too often, in their zeal to please the "customer," the marketing leaders focus their efforts on customers and neglect the needs of the eventual end users. The more remote the customer is from the end user and his or her environment, the greater is the mismatch between the product developed and what the users really needed. This is all the worse when the users reside in regions other than the one where the product is to be engineered. Features are included in the products that are not needed by the end users. Some features needed by end users are not included. The customer pays for things that are not used by end users and provide no value. Opportunities to enhance productivity are missed because some functions and features either are not included or are so inaccessible in the product, owing to poor design, that they are not used.

The global use of mobile phones and the Internet and all their associated services motivated many companies to rethink who they interview and what data they gather to understand customer needs and conceive of new products (e.g., how they conduct

Phase 1 of the product development process). Nokia has been one of the leaders in using a multidisciplinary and multicultural approach to the study of user needs and new product solutions (Mizobuchi, Ichikawa, and Shiraogawa, 2001). Early on in the mobile phone revolution, their focus shifted to the end users and the many regional cultures in which their products would be marketed. Their experience and success should motivate all companies to better understand user needs and cultural differences before they attempt to define product solutions.

Phase 1 of product development sets the stage for the future cross-cultural deployment of a product. Critical Phase 1 activities and decisions are briefly described below.

10.2.1 DETERMINE WHERE THE PRODUCT WILL BE ENGINEERED AND WHERE IT WILL BE SOLD

If the product will be developed in one cultural region, say India, and sold in another cultural region, say, the United States, then cross-cultural design will be an issue. The issue is greatly compounded if the product is to be developed in one region and sold in many other regions of the world. It multiplies the effort required in Phase 1 to understand user needs in target cultures. It drives the design philosophy and details, the plan for internationalization, or localized versions of the product. It tells you where and how you will have to evaluate the usability of your product design.

10.2.2 DESCRIBE USERS, USER GOALS, AND ENVIRONMENTS

This is the fundamental task in Phase 1 of any product development. For products to be deployed in multiple regional or cultural markets, users from each of the target regions, and perhaps even subregions, must be studied to understand their needs, goals, and environments. Chapter 12 discusses some of the problems and methods of conducting user needs research and provides some interesting examples.

10.2.3 IDENTIFY ATTITUDES ABOUT THE TECHNOLOGY

An important part of the user needs studies in Phase 1 of product development is to understand your end users' attitudes toward the technology of interest and to develop an initial model for how the product or service would have to be deployed to them in order to gain their acceptance. There is wide variation in the acceptance of technology across generations, cultures, and subcultures. The product or service will almost certainly be unsuccessful if the development team fails to understand the target users' attitudes toward new technology and their practical limitations to using it owing to issues such as illiteracy. This crucial issue determines, among other things, whether your product could be profitable given the deployment constraints posed by the end users' attitudes toward technology.

10.2.4 IDENTIFY ANY REGIONAL CERTIFICATION REQUIREMENTS

Many products are subject to a final certification by a government or national standards body to ensure that they meet national or regional standards. Examples include products in the areas of life safety, health care, consumer safety, and transportation. Increasingly, product usability is included in such certification tests conducted by these agencies. Because certification standards and testing requirements vary considerably between countries, it is best to be aware of these at the outset so plans can be made well in advance to apply sufficient resources to certification testing in-country.

10.2.5   PREPARE CROSS-CULTURAL DESIGN PLAN AND BUDGET

By the conclusion of Phase 1, the users and target markets should be identified, along with a high-level concept of the product you will make. You will be able to prepare a first draft of a plan for human factors design support during the product design and implementation. That plan must take into account all the user cultures to be served by the product. It must plan and budget for cross-cultural design, iterations on the design based on cross-cultural heuristic evaluation, and eventually usability testing with each local user group. Of course, at this early stage of product development, the plan will necessarily be just a draft, with more details and more accurate estimates filled in during subsequent phases of development. However, starting to plan for the resources needed to do a cross-cultural design will make it more likely that the resources will be available during later phases of the process to design and implement a successful product in multiple cultural regions.

10.2.6   DOCUMENTATION: USER EXPERIENCE REPORT

The cross-cultural user experience studies conducted during Phase 1 should conclude with documentation of the following:

- Identify the target markets where the product will be designed and engineered.
- Describe the users, their cultural characteristics, goals, and environments of use.
- Describe user attitudes toward the technology of interest.
- Define requirements for passing national certification requirements.
- Define early concepts for what to make and how it might look.
- Prepare an initial plan and budget for cross-cultural design, prototyping, and evaluation.

10.3   PHASE 2: CONCEPT DEFINITION

Phase 2 begins only if the studies in Phase 1 demonstrated the potential to make money from the proposed product and a decision was made to proceed to Phase 2. That said, the objective of Phase 2 of the process is to define the functional requirements for the product and the concept for what to build.

10.3.1   ESTABLISH CROSS-CULTURAL DESIGN PHILOSOPHY

Early in Phase 2, you will want to establish the cross-cultural design philosophy for the product to be developed. Two important issues to consider are the strategy for internationalizing or localizing the product and the expectations of your target users for usability.

10.3.1.1   Internationalization-Localization Strategy

Internationalization is the process of designing a product or system so that it is generic enough to accept many variations and cultural contexts and can be adapted to various languages and regions without engineering changes. Localization, on the other hand, is the process of adapting a product or system so that it can be used by people of a particular cultural context, locale, and area. Many companies perform localization by adapting a product created specifically for its domestic market. A product designed for its creator's domestic market is often embedded with the cultural markings of the creator's cultural context (Hoft, 1996). Any localization after the product is developed will require recoding and possibly reengineering to accommodate the cultural context of a target locale.

For any company that intends to extend their products for a global market, the approach of product globalization is recommended. Globalization consists of two steps. The product is first internationalized, that is, developed as a culture-free base product. This base product then is localized for each target locale. The internationalized design provides the framework and structure in which localization takes place more easily and more efficiently (Luong et al., 1995; Sun Microsystems online references). Internationalization is the preparatory stage of product development where the embedded culture and language are extracted and generalized (Taylor, 1992). The Siemens global websites are excellent examples. The same basic template is used for every localized version of the website. Localized images and textual content are simply filled in the appropriate fields.

10.3.1.2   Determining Key Usability Factors and Values

Underlying much of what is done in user-centered design is the notion that people worldwide understand the fundamental attributes of "good usability" in the same way—effectiveness, ease of use, visual appearance, efficiency, satisfaction, fun, nonfrustration (Frandsen-Thorlacius et al., 2009; ISO 9241-12, 1998). Recent research questions this assumption of universal constructs of usability (Frandsen-Thorlacious et al., 2009; Hertzum et al., 2007; Lee, 2001). Frandsen-Thorlacious et al. sampled 412 users from China and Denmark to determine how they understood and prioritized attributes of usability. Chinese users placed greater value on visual appearance, satisfaction, and fun than Danish users. Danish users valued effectiveness, efficiency, and lack of frustration more highly than Chinese users. Clearly, dimensions of usability were weighted differently for these two cultural groups. Lee (2001) studied Koreans, Japanese, and Americans and found that they differed in which elements of the user interface were most important to them. In another study, Hertzum et al. (2007) used repertory grid interviews to explore how personal constructs of usability differed between people from three different cultures, Denmark, China, and India. They found that some of the constructs verbalized by study participants were consistent with common notions of usability such as ease of use and were important at least to some degree to participants from all three cultures. Other constructs, however, differed from commonly used attributes of usability. The most important usability attributes for Chinese subjects involved issues of security, task types, training, and system issues. In contrast, Danish and Indian participants focused on more traditional aspects of usability such as "easy-to-use," "intuitive," and "liked." None of the Chinese subjects verbalized these as primary constructs of usability.

These studies suggest that knowing and clearly stating what your target users value in usability should be part of your design philosophy. Further, the design philosophy might have to accommodate different values from different cultures. In that case, the design philosophy first should emphasize the values that are common across target user groups, but drive clear plans to accommodate cultural differences in these values. From a business perspective, if you fail to consider these user values, you run the risk of emphasizing the wrong thing in your design and consuming scarce development resources to do it. One would have to question why a company would invest scarce development dollars to perfect aspects of a user interface that are not particularly important to a broad spectrum of the intended product users. Developers also should be aware that basing usability requirements for a global product on just one or two cultural groups runs the risk of minimizing attributes that turn out to be important to a second, third, fourth, nth, cultural group of users somewhere in their market space, perhaps even to the majority of potential product users. Therefore, in Phase 2, as part of the design philosophy, an attempt should be made to identify the attributes of usability that are most and least important to users in the target markets. Note that, once determined, these findings can be reused in other developments.

10.3.2 CONCEPT PREVISUALIZATION STUDIES

Previsualization prototypes provide a glimpse of what alternative renderings of the product's functionality and user interface might look like. These very early prototypes may be developed as sketches on paper or a whiteboard, cardboard mockups, wireframes in PowerPoint, or a combination of media integrated via Flash. They show the functionality available to the user and some preliminary and possibly alternative notions of screen layouts and navigation. Cardboard or foam-core mockups show users alternative form factors for the product concept. Previsualization also provides an opportunity to define a consistent model of high-level screen layout and interaction that will ensure a and unified user experience. Whatever of the above forms they take, these early prototypes are sufficient for the user to review and to determine, at some relatively high level, if this is what the user had in mind when he or she described their needs to the human factors researcher in Phase 1.

10.3.3 USER REVIEWS OF CONCEPTS IN TARGET CULTURES

Once previsualization prototypes are developed, there is no substitute for reviewing them with local people from the targeted cultures. Be mindful of the potential problems involved in your being a developer from one culture attempting to obtain feedback from users in another culture. Chapter 13 provides an overview of these design evaluation problems and guidance about how best to conduct them.

10.3.4 DOCUMENTATION: USER INTERFACE CONCEPT DOCUMENT

- Describe the user interface design philosophy including globalization and usability strategies that will lead to efficient development of appropriate localized versions of the product.
- Provide previsualization prototypes of the user interface to illustrate functionality and interaction concepts.
- Report on any reviews of the previsualization materials with customers and end users.

10.4 PHASE 3: DESIGN AND SPECIFY PRODUCT

The purpose of Phase 3 is to convert the design concept and functional requirements developed in Phase 2 into detailed design requirements and to produce a development schedule for Phases 4 and 5. The most critical aspects of Phase 3 are insuring that the requirements are well defined and understood by all members of the development team. If part of your development team is located in China or India, this shared understanding of requirements and their context must include them (Plocher and Jin, 2008).

10.4.1 DESIGN, DEVELOP, AND ITERATE USER INTERFACE (UI) PROTOTYPES ACCORDING TO GUIDELINES

During Phase 3 the cross-cultural design guidelines described in Section II of this book can be applied in all their detail as the concepts and functional requirements from Phase 2 are translated into a detailed design. It is the time to conduct studies of information architecture to design the organization of information and navigation using a tool such the Card Sort Analyzer of Deibel, Anderson, and Anderson (2005). Prototype screen layouts can be designed with careful attention to language, directionality, and placement of information. Visual design elements, including appropriate metaphors, icons, and imagery, can be selected. These initial prototypes should be tested in the target cultures, revised, and tested again.

10.4.2 DESIGN FOR TECHNOLOGY ACCEPTANCE

A number of things can be done during design to increase the likelihood of product acceptance by target users:

- Make it easy to use.
  - Avoid using too many characters on the interface; use appropriate symbols and icons instead.
  - Put significant hotkeys on the main interface.
  - Provide simple but effective help and demonstration.
- Facilitate positively on users' social impression: Study the social impression requirements of different ages. For example, the younger generation

seldom accepts a product that looks out of date, because it may result in an impression by others of its user "being out of date."

- Help social connection in collectivistic cultures: Involve interpersonal invitation and recommendation functions.
  - Let users know who else is using it and how many other users there are.
  - Recommend related products through social computing; for example, "people who play game A also play game B."
- Focus on information accessing in individualistic cultures.
- Show a reliable quality: Be clear, readable, responsive, and attractive enough to show a reliable quality, especially for the first impression.

10.4.3 CROSS-CULTURAL HEURISTIC EVALUATION

Heuristic evaluation is a cost-effective way to track the evolution of the UI design and keep track of design issues and their resolution. Prototypes can be evaluated using heuristic criteria iteratively throughout the detailed design phase to provide a measure of progress during each step toward the final design. Chapter 14 describes how to incorporate cross-cultural design evaluation criteria into a more traditional heuristic evaluation methodology. Cross-cultural heuristic evaluation is no substitute for usability testing with real users in the target markets for the product. However, it can be an effective guide toward iterative design improvements. It can be particularly effective if the human factors experts conducting the cross-cultural heuristic evaluation are selected from the target cultures.

10.4.4 USABILITY TESTING IN TARGET REGIONS

In Phase 3, when user interface prototypes are sufficiently mature and provide sufficient user interactivity, it is time to gather user feedback via formal usability tests in the target markets. Chapter 13 discusses many of the pitfalls and problems involved in testing user interface prototypes in other cultures, as well as solutions and guidelines to ensure your investment in regional or local usability testing pays off.

10.4.5 DOCUMENTATION: HUMAN FACTORS DETAILED DESIGN DOCUMENT

The detailed user interface design developed in Phase 3 will document the following:

- Design of each screen or display, including layout and visual design
- Navigation between screens
- Storyboards to illustrate interaction with the user interface to perform the required tasks and scenarios

10.4.6 DOCUMENTATION: HUMAN FACTORS EVALUATION REPORT

Although shown in Figure 10.1 as originating during Phase 3, the evaluation plans and methods described in this report may actually be documented much earlier in the process, during concept development. Some early evaluations may be conducted in Phase 2, as well. That said, testing and evaluation will be more frequent and more comprehensive as the detailed design evolves during Phase 3. The findings and recommendations are documented and fed back into the design process.

10.5　PHASE 4: DEVELOP HARDWARE OR SOFTWARE

For software products, implementation or coding takes place in Phase 4. The task for the UI designer shifts from designing the user interface to monitoring and evaluating successive versions of the software to ensure that they faithfully implement the overall UI design philosophy and the detailed design defined in Phases 2 and 3. In this world of software engineering by groups in offshore, low-cost centers, this can be a challenge. In many corporations, the people who code the software for a new product are not the same people who studied the end users, conceived the product, wrote the requirements, and developed the detailed design. The people who end up coding the software may have only the detailed design specification from which to work. User interface designers must play a key role in filling this information gap by providing important contextual information to the developers. They can do this by explaining the context of the product to the developers and expanding on the deliverables prepared during Phases 1 and 2 of the development process. They must also guard against drift in the detailed design and the overall design philosophy. Both cross-cultural heuristic evaluation and usability testing in the target markets will help validate that the implementation has followed the intent.

The design and implementation of hardware products tends to proceed more quickly than that of software products. The design is frozen early in Phase 4 or even late in Phase 3, after which changes, including usability-driven changes, are difficult and costly to make. Examples of such products are protective equipment and clothing, handheld devices, and body-worn sensors. All are products in which the ergonomic design and anthropometry are key discriminators that must be tested and finalized before molds and dies are made in Phase 4. Any cross-cultural issues related to comfort and fit, regional anthropometry, and ergonomics must be identified and incorporated into the product design by the end of Phase 3 or early in Phase 4. A revised Human Factors Evaluation Report should be produced to document any evaluations done during the product implementation.

10.6　PHASE 5: PRODUCT VALIDATION

In Phase 5, the performance of the product is validated against the requirements and design specification. If the product must be certified by a government or trade organization, that certification testing is done at this time. Cross-cultural issues potentially arise when the product must be certified in a country other than the one in which it was developed. It can be difficult to recruit the required human factors expertise in that destination country to support the certification tests. This is particularly problematic in a highly technical domain such as aviation, where certification testing requires both aviation domain expertise and human factors expertise. Considering certification testing requirements early in product development should allow sufficient time for planning how you are going to meet certification testing support needs.

A final iteration of the Human Factors Evaluation Document should be produced to document any discrepancies between product specification and product implementation.

10.7 PHASE 6: PRODUCT LAUNCH AND DEPLOYMENT

10.7.1 IMPLEMENT TECHNOLOGY ACCEPTANCE/DEPLOYMENT PLAN

The plan for product roll-out and technology acceptance might include one or more of the following:

- Explain the possible cost clearly when users first see the product: Show that the product does not have so many cost risks as the user might imagine. An example saying could be "it won't cause any additional charges."
- Provide users with trials before they decide to pay for it.
- In developing countries, deploy product through community or family organizations, which can provide an environment for mentoring and training.
- Encourage users to feel at ease when introducing a product: Saying "just press one key" or "don't worry about any bad consequence," and so forth.

10.7.2 COLLECT USER FEEDBACK

Once a product is launched into the marketplace, the responsibility for it shifts to marketing and sales, customer service, and sustaining engineering. In most manufacturing companies, none of these has a strong connection back to the human factors engineering team. User feedback from the field may be collected by the sales team but may not find its way back to the human factors team. The opportunity to either improve the cross-cultural usability of the current product through new releases or to enhance future products is lost. Deploying a formal mechanism to achieve that flow of feedback from the field to the human factors team is one solution. Better yet is to involve the human factors team directly in soliciting feedback from the field.

10.7.3 ADD PROJECT TO CULTURAL DATABASE

Much of the knowledge acquired during the project is reusable. This includes knowledge about the target cultures, users, and local environments, as well as user interface design elements that were vetted via usability testing with local users. However, it is reusable only if it is well documented in some form, such as an online team room, that is readily available to future engineering projects.

REFERENCES

Frandsen-Thorlacius, O., Hornbæk, K., Hertzum, M., and Clemmensen, T. 2009. Non-universal usability?: A survey of how usability is understood by Chinese and Danish users. CHI 2009: Digital life, new world : conference proceedings and extended abstracts; the 27th Annual CHI Conference on Human Factors in Computing Systems, April 4-9, 2009 in Boston, USA.

Hertzum, M., Clemmensen, T., Hornbæk, K., Kumar, J., Shi, Q., and Yammiyavar, P. 2007. Usability constructs: A cross-cultural study of how users and developers experience their use of information systems. *Lecture Notes in Computer Science*, 4559: 317–327.

Hoft, N. 1996. Developing a cultural model. In *International User Interfaces,* ed. E. delGaldo and J. Nielsen, 41–71. New York: Wiley Computer.

ISO 9241-12: 1998 (E). *Ergonomic requirements for office work with visual display terminals (VDTs)—Part 12: Presentation of information.*

Lee, K.P. 2001. Culture and Its Effects on Human Interaction with Design: With an Emphasis on Cross-Cultural Perspectives between Korea and Japan. PhD dissertation, Institute of Art and Design, University of Tsukuba, Japan.

Luong, T.V., Lok, J.S.H., Taylor, D.J., and Driscoll, K. 1995. *Internationalization: Developing Software for Global Markets.* New York: Wiley.

Mizobuchi, S., Ichikawa, F., and Shiraogawa, A. 2001. Localization of a user interface for Japanese market: Nokia's challenges in mobile terminals. HCI International, Ninth Conference on Human-Computer Interaction, New Orleans, LA, August 9.

Plocher, T., and Jin, Z.X.J. 2008. Working together as a global human factors team: American and Chinese perspectives. In *Proceedings of the Second International Conference on Applied Human Factors and Ergonomics.* Las Vegas, NV, July 14–17.

Taylor, D. 1992. *Global Software: Developing Applications for the International Market.* New York: Springer-Verlag.

11 User Needs Research
Understanding the Lay of the Land

11.1 INTRODUCTION

The earliest phase of product development is aimed at answering the questions of what to build and who will buy it. Traditional approaches to answering these questions tend to be based on the notion of extending current products by adding new functionality (Kumar and Whitney, 2007). Thus, the research to define the concept and market for the new product focuses on how customers react to current products or preconceived prototypes. Surveys, focus groups, and interviews are commonly used to identify problems with the current product and improvements that the customer would like to see. There is nothing particularly wrong with this approach if your goal is to define an incremental extension of an existing product in a tactical manner. But it artificially constrains the scope of the user needs investigated to those that are closely associated with the legacy product and thus were previously expressed to developers. It constrains our knowledge of the context of product use to those contexts that the customers remember to report in the survey, interview, or focus group session. The result is that the knowledge of the user and product which we gain from this exercise is detached from the natural phenomena of the users' lives. Unmet user needs are missed if they lie in the space adjacent to the need met by the existing product. Completely new needs within the space go unstated. It is not surprising that this traditional market-oriented approach rarely leads to insights that translate into highly innovative extensions of the legacy products or altogether new products in adjacent or completely new spaces.

In contrast to more traditional approaches discussed above, user needs research focuses on describing the users, their cultural characteristics, their unmet needs, their goals, and their environments of use. The synthesis of these data frequently results in new and innovative concepts for products and product features and functions.

A variety of methods are used, including surveys and interviews. However, the key to this approach is that those traditional methods are almost always combined with a significant amount of firsthand observation or recording of the details of how the target users live and perform tasks in their natural work or home environments.

Effective user needs research has to deal with a number of problems. Adding a cross-cultural dimension to it—that is, people in one culture designing a product for people in another culture—creates additional problems.

- The methodology has to be discovery oriented and not overly constrained by the business or product biases of the researchers.
- The methodology must be capable of capturing a detailed record of the context in which future products would be used by the target population.
- It must be affordable and within the project budget, which may be tight.
- The methodology must be able to reflect users' acceptance for new technologies in a particular cultural context.
- Particularly in Asian cultures, the methodology must overcome the subject's natural hesitation to talk freely and directly about the subjects of interest.
- The methodology must serve to neutralize any ethnocentric biases held by the researchers.

11.2 REDUCING ETHNOCENTRICITY IN CROSS-CULTURAL RESEARCH

Li, Yang, and Wen (1985) expressed the concern of many psychologists and social scientists about using a Western cross-cultural approach to understand non-Western people. In addition, Yang (2001a) collected her previous published papers on how to study the Chinese, the indigenous approach. She systematically summarized the ways to conduct studies in China and how to localize studies in China. The indigenous approach Yang adopted is based on local materials and observations, a set of commonly shared meaning systems with which the people under investigation make sense of their lives and their experiences, and present and derive meanings while interacting with each other (Yang, 1991, 2000a, 2001a, 2001b). This also helps indigenous researchers understand and interpret the behaviors manifested by the people under study.

Bond (1986) wrote extensively about the psychology of the Chinese people, discussing Chinese patterns of socialization, perceptual processes, cognition, personality traits, psychopathology, social psychology, and organizational behavior. This now classic book provides great insights about psychological and cultural differences between Chinese people and people from other cultures.

When conducting cross-cultural studies in China, special issues concerning cross-cultural comparison should be carefully considered to ensure the reliability and validity of the study. These lessons from research in China apply to other cultures of the world as well.

11.3 COUNTRIES ARE NOT NECESSARILY PROXIES FOR CULTURES

Researchers often use country as a proxy for culture. For example, they select participants from China and the United States to represent Eastern and Western cultures. Schaffer and Riordan (2003) found that as many as 79% of the cross-cultural organizational studies in the literature used country as a proxy for culture. That is a somewhat naïve practice.

Consider, for example, the incredible diversity of "Chinese" users. Chinese users include people in Mainland China, Taiwan, Hong Kong, Singapore, Malaysia,

Thailand, and other parts of Asia. Mandarin is the main spoken dialect among Chinese in Asia. However, ISO 639-3 (SIL International, 2011) identifies an additional 12 dialects that are spoken regionally. Most of these are, to various degrees, mutually not intelligible. Among these, Wu or Shanghainese, Yue or Cantonese, Min or Hokkien each are spoken by some 50 to 100 million Chinese. Writing systems for Chinese language also differ around Asia, with simplified Chinese used in Mainland China and traditional Chinese used in Taiwan. Not surprisingly, even within the population in Asia with Chinese ethnicity, there are diverse subcultural groups, as well.

For example, within the past 50 years, Chinese users in Taiwan have had more opportunities to learn American culture than Chinese users in Mainland China. Thus, differences between users in Taiwan and users in the United States tend to be smaller than the differences between users in Mainland China and users in the United States (Rau, Choong, and Salvendy, 2004). Cultural differences also exist between urban and rural Chinese and, of course, between the generations of people. Plocher et al. (2001) found that the Tsinghua University students who participated in their study of value orientation did not reflect the collectivist values that researchers tend to associate with traditional Chinese culture.

11.4 MEASUREMENTS ARE NOT NECESSARILY EQUIVALENT

Questionnaires are a part of many cross-cultural user research studies. Often the questions are conceived in one language, such as English, and translated into the language of the target user. The ability of the questionnaire to measure what the researcher intends is highly dependent on the quality of that translation as well as the equivalence of concepts underlying the questions. Brislin's (1970) classic paper investigated the factors that affect translation quality and how equivalence between source and the target versions can be evaluated using back-translation.

11.5 CONCEPTS ARE NOT NECESSARILY EQUIVALENT

It is easy to assume that other cultures share concepts equivalent to those developed in the United States. After all, the United States has been a world leader in academic research, and many of the concepts in academia around the world originated in U.S. academic institutions. The influence of American popular culture, particularly through entertainment, in other parts of the world is also well known. Despite all these influences, both academic and popular, user research must be aware that some of the concepts developed and promoted in the West may not necessarily have equivalent concepts in the East.

Consider, for example, the concept of personality traits. The concepts of self-construal, achievement, and so forth, may have different meanings in China. Farh, Zhong, and Organ (2004) compared the forms of organizational citizenship behavior (OCB) that appeared in Chinese and Western literature. They found that 10 dimensions of OCB were commonly described in China, with at least one dimension not evident at all in the Western literature. Consider also the concepts underlying medicine. Widely different concepts of the human body, health, disease, and treatment underlie the allopathic medicine of the West and traditional Chinese medicine in the East. There are no equivalents in China for many of the concepts of allopathic medicine and no equivalents in the West for many concepts of traditional Chinese medicine. Thus, it is crucial that the concepts underlying the subject of a particular user study are clearly understood and articulated in the same way both by the user participants in the study and the user needs researcher. Validating any assumption of concept equivalence is an important part of the study planning process.

11.6 CASE STUDIES

This chapter does not intend to provide a complete review of methodologies for user needs research. We prefer to recommend a good source reference on field methodology for user needs research, such as Randall, Harper, and Rouncefield (2007). That said, we encountered a number of fascinating approaches for discovering user needs that support ideation of product concepts while effectively addressing the above problems. We present two projects below as case studies.

11.6.1 CASE STUDY: USING PHOTOINTERVIEWS TO UNDERSTAND INDEPENDENT LIVING NEEDS OF ELDERLY CHINESE

Plocher and Zhao (2002) applied the photointerview method of Collier and Collier (1999) to a study of the needs of elderly Chinese for home security, environmental comfort, communication, and other aspects of independent living. The goal was to understand these needs and how they translated into product concepts, requirements, and constraints.

The photointerview method developed by Collier and Collier (1999) is a hybrid between visual ethnographic and more conventional structured interview techniques. It is economical in that it does not require hundreds of hours of labor to analyze video recordings. The photointerview method involves collecting a complete and structured series of still photographs in the subject's environment and then using those photographs to elicit comments and stories from the subject during a structured interview. The key is that the structure of the interview is provided by the photographs (e.g., the context in which the subject lives), and not simply by the questions brought to the session by an outside interviewer. Kumar (2007) has written extensively about this methodology. Successfully, he gave disposable cameras to the participants and had them, rather than the researchers, make the photographs.

11.6.1.1 Methodology

11.6.1.1.1 Participants

Four elderly Chinese couples participated in the study, two from Beijing and two from Shanghai. The average age of these eight participants was 72 years and ranged from 62 to 78 years. Three of the four families were from similar socioeconomic circumstances, two being families of retired professors and the third the family of a retired joint venture company manager. These three couples were typical of the new Chinese middle class. The husband of the fourth couple was a retired textile worker and reflected a slightly lower economic situation than the other three.

11.6.1.1.2 Interviewers

The interview team consisted of one American from a corporation in the United States and one lead Chinese interviewer from the Chinese Academy of Sciences Institute of Psychology. A third team member attended the sessions as a data recorder. The interviews were conducted in Chinese.

11.6.1.1.3 Field Data Collection Protocol and Materials

11.6.1.1.3.1 Visit Schedule Two days were needed for the photointerview visits. The first day of the visit served to introduce the elderly couples and the researchers, usually over tea. The elderly couple was asked to lead the researchers on a walking tour of their extended neighborhood, apartment complex, apartment building, and apartment and point out things that they thought were important or special. During the tour, a complete and structured series of photographs was made of the home and the neighborhood. The photographs then were taken to a shop for processing. The final task on the first day was for the researchers to sort through all the processed photographs (usually a hundred or more) and select and organize those that would be used in the interview on the second day.

On the second day, the researchers returned to the elderly couple's home and conducted the interview. The photographs were organized by living area and spread out as a collage on a table in the home. Everyone sat around the table and looked at the photographs. During the interview, the researchers let the photographs serve as points of departure for discussion and stories. When necessary, the researchers prompted discussions with open-ended, unbiased questions such as "What's this?" or "Can you tell me about that?" (pointing to something in a photograph). Only when it appeared that a topic would be completely missed in the discussion did the interviewer ask a specific prepared probe question.

11.6.1.1.3.2 Structured Checklist for Photographs If the same series of photographs is recorded at each subject site, then, during analysis, the photographs can be compared, both qualitatively and quantitatively across subjects. The photographs become the basis for an object inventory, the results of which can be reported in descriptive statistics across subjects. To aid in replicating the same photographic scenes from site to site, Plocher and Zhao (2002) prepared a checklist of standard scenes to photograph. The checklist, a sample of which is shown in Figure 11.1, was structured around the various living spaces in the typical Chinese urban neighborhood, shown in Figure 11.2.

Major categories included the following:

1. The extended neighborhood: The outermost portion shown in Figure 11.2 which included the city streets bounding the apartment complex and the shops immediately outside the walls or fence of the complex.
2. The inner neighborhood: Enclosed within the apartment complex wall is the inner neighborhood consisting of 25 to 50 apartment buildings, streets, alleyways, usually a garden and exercise area, and an occasional street vendor.

Scene	Daytime	Nighttime
Panorama of balcony standing at entrance		
Wall 1		
Wall 2		
Wall 3		
Wall 4		
Ceiling		
Windows		
Doors		
Close-ups of objects		
People		

FIGURE 11.1   Sample photograph plan.

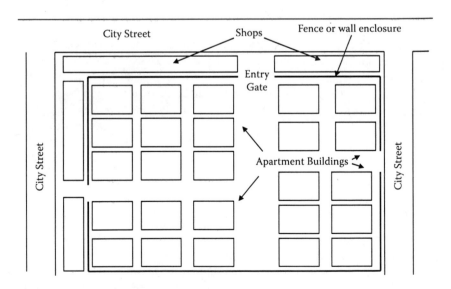

FIGURE 11.2   Typical urban Chinese neighborhood.

3. The apartment building: Each building typically had common areas including an entryway in the front of the building, a lobby with mailboxes, and a stairway or elevator to upper floors of apartments.
4. The apartment or flat: The typical apartment was composed of a dining area, cooking area, living room, one or two bedrooms, water closet, and a balcony.

Within each of these four categories, the photographic checklist contained a list of objects, features, and activities to record if the opportunity presented itself.

11.6.1.1.3.3 Structured Interview Questions An interview guide was developed and consisted of two parts. One part was the "reminder" table (Figure 11.3) consisting of the high-level user needs (comfort, convenience, health, safety, and social well-being), and the user activity variables (activity, environment, interactions, objects, users), of interest to the study. This short list was always available to the interviewer and served as a constant reminder about the variables of interest that needed to be discussed. When the interview conversation lagged or drifted away from the intended focus, these variables provided a way to quickly improvise a question and reset the focus. Such impromptu questions were often framed in terms of the interaction between two or more of the variables and needs. For example, "Is it comfortable (need) for your grandson (user) to sleep (activity) in the living room in the summertime (environment)?"

The second part of the interview guide was a series of probe questions organized around the same four categories of living space that provided the photographic framework: extended neighborhood, inner neighborhood, apartment building, and apartment. The sample in Figure 11.4 focuses on "the balcony," a common space in the typical Chinese home noted for its adaptability to a wide range of activities. The interview guide was used primarily as a checklist to record topics covered in the discussion. When a topic failed to be raised and discussed spontaneously, the questions were used directly as probes.

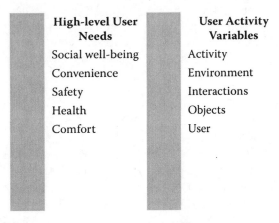

High-level User Needs	User Activity Variables
Social well-being	Activity
Convenience	Environment
Safety	Interactions
Health	Objects
Comfort	User

FIGURE 11.3 "Reminder" chart aided the interviewer. (Activity variables adopted from the Doblin consultancy.)

Balcony

Consider your activities in the balcony area. /What things do you do in the balcony? How frequent is each activity?

Do you use the balcony as part of living room or kitchen, or anything else? Why? If it's possible, do you like to change it? How?

FIGURE 11.4 Example of structured interview questions to support photointerview.

11.6.1.1.3.4 Photointerview Products The following products were produced by the photointerviews:

1. Systematic and comparable photographic records of the living environments of the four families were used to create an object inventory from which summary statistics were compiled. The photographs also provided the context essential for interpreting the subject's interview comments.
2. Interview responses from the four families which were compiled and interpreted in the context of the photographs to draw some conclusions about activities of daily life, concerns, and values.
3. The photographs served as an archive that was used to answer specific post hoc questions about details of the living environment and how home products might be designed to fit the environment.

11.6.1.2 Findings

11.6.1.2.1 Object Inventory

Results of the object inventory are shown in Figures 11.5 and 11.6 in terms of what objects people did and did not have in their homes. The number of families

---

Air conditioners (3/4)	Window security bars (3/4)
Two or more cooling fans (4/4)	Metal security door (3/4)
Ceiling fans (2/4)	Bottled water (4/4)
Infrared bathroom heater (3/4)	Electronic kitchen appliances
Home computer (3/4)	(4/4-lots)
Telephones (4/4)	Gas hot water heater (4/4)
Cell phone (1/4)	Gas stove (4/4)
Two TVs (3/4)	CO sensor/alarm (1/4)
Entertainment center/stereo	Washing machine (4/4)
system (3/4)	Chinese/Japanese brand name
Plants (4/4)	appliances (4/4)
Exercise equipment (3/4)	

FIGURE 11.5 Objects in homes: what people had (expressed as the number of families out of four that possessed the object).

Fire and smoke alarms	Water filters
Electronic window and door	Clothes dryers
intrusion alarms	Controllable central heating
Building access security	Indoor plant lights
Complex access security	American brand products
Air cleaners	

FIGURE 11.6 Objects in homes: what people did not have. (None of the four families possessed these objects.)

possessing the object out of the four couples studied is shown. Interestingly, three of the four families had home computers, and all the families had a wide range of electronic appliances and home entertainment equipment. Everyone grew plants in the home. Security was accomplished by physical barriers (locks, doors, and bars) rather than by electronic devices. Drinking water was from a bottled water source rather than from a filter device. Air cleaners and fire/smoke alarming devices were absent. All objects were Japanese or Chinese made. To the American investigator, some surprising findings emerged from the object inventory. These included the prevalence of computer technology in the home and the omnipresence of various electronic devices and appliances. Just as striking was the lack of technology for fire safety, security, and water purification.

11.6.1.2.2 Independent Living Values and Concerns

Comments and stories from the participants were reviewed in the context of the object inventory and clustered together around common themes to better understand their independent living problems and needs. These are summarized in Table 11.1, along with the larger political and social trends to which they are related. The researchers were aware of some of these trends prior to the study. Other trends were pointed out and elaborated by participants during the interviews.

The most frequent concerns expressed by the elderly subjects during the interviews were those related to the importance of good health and how to maintain it by means of physical activity rather than medication. Also frequently mentioned was the need to improve physical security in their building and complex. One couple raised the issue of social isolation as an increasing concern among elderly Chinese, and the need for a way to stay connected with old friends and family. Comfort and convenience, which usually rank high among independent living concerns in the United States, were rarely mentioned. High value was placed on the natural environment as a source of pleasure. Further, their idea of "comfort" was associated with natural cooling, by means of open windows and fans, rather than artificial air conditioning. One of the things the researchers noticed during their first day tour of these homes was the presence of both a modern air conditioner and two or more cooling fans, sometimes even in addition to a ceiling fan. Why both? Discussions with the elderly people revealed that even though they owned an air conditioner, they rarely used it. Rather they preferred the more natural cooling and flow of air provided by

TABLE 11.1

Political-Social Trends in China, Problems of the Elderly and Solutions

Trend	Problem	Potential Solutions
Government is selling off its housing and phasing out housing assignments based on work unit	Formerly, people from the same work unit lived together in the same complex well into their old age. Everyone in the neighborhood knew everyone else. People are now moving into housing based on convenience and availability. The result is an increasing number of elderly people isolated from friends in an unfamiliar neighborhood.	Personal security and safety; personal communications
Increasing personal crime	People feel very safe walking around their neighborhoods. But apartment break-ins are a common problem. Building security relies on physical barriers (bars and doors). Complex security relies on low-paid security guards who are inattentive.	Better apartment security particularly from window entries
Increasing concern about fire safety due to well-publicized and devastating recent fires (10,000 fires in 1 year in Liaoning Province, mostly in homes and killing 89 people)	With all their windows protected by bars (to keep the criminals out), people have only one route of egress in case of fire (e.g., their front door and then the front door to their building). None of the apartments or buildings we saw had smoke detectors, alarms, sprinkler systems, or fire extinguishers.	Need for better smoke and fire warning systems in individual apartments and in buildings
Government is attempting to reduce the service bottlenecks at medical facilities	Clinics are overburdened with patients, particularly elderly people, causing service bottlenecks. Also, it is difficult for many elderly people to travel to a clinic (foot, bike, bus). Many of these patients have symptoms that do not warrant direct medical attention.	Improved system for telephone-based assessment of symptoms, screening patients, and medical advising
Elderly population of 132 million is growing by 3.2% each year. By 2040, elderly will account for 26% (400 million) of the population. In major cities, they will make up 33% to 50% of the population.	Increasingly, young Chinese are changing from traditional values. They are less directly involved in their parents' and grandparents' lives and well-being.	Some substitute, such as remote monitoring, for the health, safety, and security provided in the past by their extended family

open windows together with fans. So they keep their windows open all night long. This, in spite of the fact that they had air conditioning.

In a seeming contradiction to this, they had a growing concern about security. Thieves were known to climb the walls of apartment buildings and enter through open windows. So what kind of technology would meet their security needs? Because the windows were always open in the summertime, glass breakage sensors would be of little use. On the other hand, a security system using motion detectors would provide some protection. The photos also revealed that the electrical junction boxes for each apartment were located in a public area, the elevator lobby. All an intruder needed to do to disable the power in an apartment was to clip the wires arising in this junction box near the elevator. Any security system that relied on the delivery of power over a wire would be instantly disabled. The product lesson from this was that any home security system would have to be battery powered and wireless.

11.6.2   Case Study: Health-Care User Needs and Concepts in Rural India

11.6.2.1   Introduction

Health care in rural India is characterized by a shortage of skilled medical providers at the point of need in the local villages. Typically, villagers have to travel to a clinic in the nearest town for skilled medical care. The providers in these clinics are overwhelmed by the number of patients, resulting in long waiting times for service. The distribution of medications, health products, and health-care information is limited in these villages.

The case study described below was aimed at understanding the rural health-care situation in India and conceiving products and services that could have an impact on the current state of rural health care (Pulik et al., 2007). The project was conducted under two initiatives, one a collaboration between Honeywell Technology Solutions–Bangalore, Illinois Institute of Technology, and the Indian Institute of Technology–Mumbai, and the other a collaboration between Honeywell and MART, a rural research agency in India. The overall approach to the project is shown in Figure 11.7 and includes phases of research and analysis, brainstorming, and ideation. The case study presented here focuses on the first phase of their research,

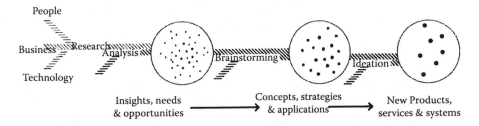

FIGURE 11.7   Overview of process for understanding user needs, brainstorming, and ideation to identify new products, services, and systems (Pulik, L., Bairagi, A., Sardana, A., Trivedi, P., Merchant, S., Gaswami, N., and Ganguly, S. 2007. Creative research as a tool to inform and inspire. *The Exponent*, May 3–7.)

aimed at understanding the overall health-care system in rural communities and its relationship to unmet user needs.

11.6.2.2 Methodology

11.6.2.2.1 Participants

The first phase of study, aimed at understanding and documenting the rural health-care system and scenarios, and identifying need gaps, was conducted in two states, Uttar Pradesh and Tamil Nadu. Uttar Pradesh in the North of India is known as a low health infrastructure state, and Tamil Nadu in the South is known as a high health infrastructure state. Within each state, two districts were identified for study: one with low infrastructure and one with high infrastructure. One town or district headquarters and two large (population > 5,000) villages were visited in each district. Thus, studies were conducted in a total of 12 villages, divided about evenly between those that had a Primary Health Center (PHC) and those that did not.

Three segments of the population, summarized in Table 11.2, participated in the study:

- Health-care consumers (drawn from upper socioeconomic classes)
- Health-care providers (drawn from both public and private sectors)
- Key influencers (nongovernmental organizations [NGOs] and ANMs)

11.6.2.2.2 Field Data Collection Approach

One of the challenges in this research was understanding rural health care from the individual villager and provider points of view, and understanding it as an overall system in the community. Further, it was important to create interview environments in which participants felt comfortable to tell their stories, engage in the issues of interest, and reveal unarticulated needs. Thus, a carefully coordinated set of methodologies was used during the site visits.

11.6.2.2.2.1 Map Making   A map-making exercise was conducted as a way to engage villagers in a discussion of the health-care system with other villagers and with the researchers. Participants were asked to draw a map of their village and

TABLE 11.2
Numbers and Distribution of Study Participants

Participant	Research Method	Primary Health Center (PHC) Village	Non-PHC Village	Town	Sample per State	Total
Male consumer	Interview	5	5	0	10	20
Male consumer	Participatory activity	1	1	0	2	4
Female consumer	Focus group	1	1	0	2	4
Health-care provider	Interview	2	1	13	15	30
KOL	Interview	1	1	2	4	8

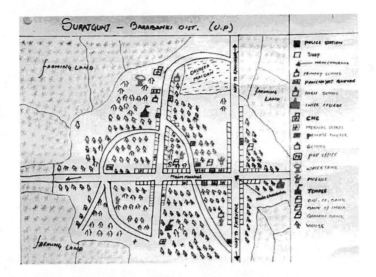

FIGURE 11.8 Map of Suratganj drawn by villagers showing important features and facilities in relation to health-care facilities and functions (**see color insert**).

annotate items related generally to their lives and how the village functioned, and particularly to health care. From the map, the socioeconomic structure of the village and the key influencers could be inferred. The distribution of health-care elements around the village was described. Most importantly, the map-making exercise stimulated discussion among participants who might otherwise have been reluctant to participate. It introduced discussion topics that could be further probed by the researchers. An example of a map produced by a group in the village of Suratganj is shown in Figure 11.8.

11.6.2.2.2.2 Workflow Mapping   Direct observations in health-care facilities and in-depth interviews with villagers and providers helped to map out the typical patient–provider workflow within and between health-care facilities. Interviews with villagers revealed the typical activities, delays, and costs encountered when seeking treatment for illnesses of varying severity, from home treatments to the local medical practitioners to a local Primary Health Center to a Community or District Health Center. The workflows revealed the costs and delays associated with the overall health-care system and validated participant comments from the interviews. The workflow maps highlighted that, ironically, the sometimes long circuitous route toward eventual treatment by a doctor in the city, with many stops at facilities or lower-level practitioners in between, actually costs more than going directly to a doctor. Figure 11.9 describes one sample workflow map of the activities, delays, and costs of a patient seeking treatment for a minor illness. Figure 11.10 illustrates the mapping of treatment routes typically followed by women, children, and older adults for treatment of nonemergency conditions.

11.6.2.2.2.3 Valuating Health Care   Indian villagers are often reluctant to discuss personal finances in terms of real currency expenditures. The researchers

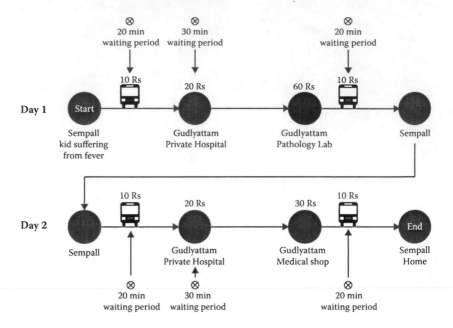

FIGURE 11.9 Sample workflow map of the activities, delays, and costs of a patient seeking treatment for a minor illness.

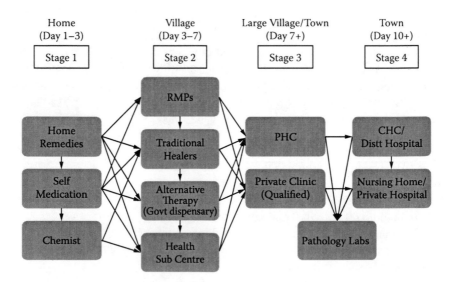

FIGURE 11.10 Mapping of treatment routes typically followed by women, children, and older adults for treatment of nonemergency conditions.

quickly found it almost impossible to accurately determine health-care expenditures from the interviews.

Two methods were used to gain a more accurate picture of personal finance in the village. In the first method, villagers were given stacks of tokens with no specific monetary association. They placed numbers of tokens on various categories of personal expenses, including health care, to indicate their relative expenditures. The activity was done in groups to elicit discussion about personal finances in the village. In a second method, a health-care valuation tool was developed as a way to understand the actual, rather than perceived, monetary value that local consumers placed on various health-care products and services in the community and relatively how much they were spending on items. The Value Equation Tool allowed participants to rate their relative, rather than absolute expenditures on health care by simply moving a slidebar on a pricelist-like chart next to each health-care product or service. Figure 11.11 shows the tool in use.

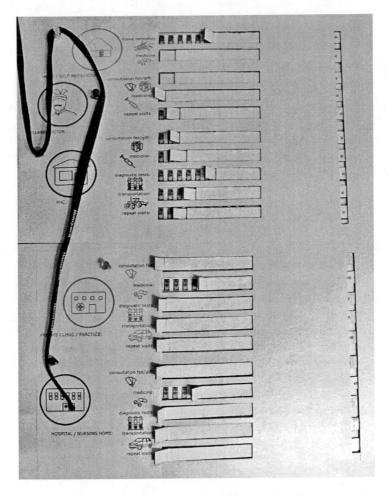

FIGURE 11.11 Participants used a Value Equation Tool for estimating their health-care expenditures (**see color insert**).

1	location/person	observation/story	Insight
2	PRA, primary school	a common belief that disease was a part of life *"eeshwar ki dein hai bimari"*, one villager	
3	PRA, primary school	*"jab chakki chalani padti thi, tub bache ghar pur hote they"*, a village senior (a belief that the women were more strong earlier to bear the pain)	
4	PRA, primary school	Since there has been less rains agricultural base has been hit and was the top concern throughout the discussion	Are there any health issues which are caused due to lack of humidity in the air? Need to probe
5	PRA, primary school	there were general misconceptions about diseases like TB, HIV and high /low blood pressure, though these things have been heard about	Awareness is clearly the first needed step, even before prevention can be thought of
6	PRA, primary school	one of them suggested, first tell me if I am sick or not then tell me what to do	
7	PRA, primary school	the common belief (which was validated later with other villagers too) was that the root of all problems lay in the use of fertilizers in the soil	
8	PRA, primary school	another common belief (mostly correct) was that allopathic mediceine only provides an immediate relief	
9	PRA, primary school	some villagers were also open to the use of homemade medicine (neem kaadha) as the second measure, i.e. if the allopathic medicine is not able to provide immediate relief (the third measure would be visiting jhansi)	
10	PRA, primary school	*"local doctor hamare saath experiment karte hain"*, since the doctors at the village were not really qualified and always refer the patients to other doctors, the faith level in them is quite low	good opportunity as the doctors also would want to redeem themselves
11	PRA, primary school	a lot of local alcohol consumed by the villagers	implications need to be studied
12	PRA, primary school	due to the lack of rains, all poor villagers could not afford to stay back in the village (approx 60%, left for citites to work at constructions sites and other such work). The villagers who stayed back are the ones who either have a service (govt jobs like teaching, grameen bank etc)	
13	PRA, primary school	**malaria, piles, gastric trouble, flu,delivery problems, cancer, TB ------ major illnesses**	

FIGURE 11.12   Sample record of field observations.

11.6.2.2.2.4   Image Mapping   Photographs were taken during observations in the villages. The photographs then were used in an image mapping activity during interviews with participants. Categories of information related to the health-care system, such as doctor, hospital, transport, and medicine, were laid out on a large paper. The participant placed photographs on the categories with which they associated them. The activity stimulated stories from the participant and provided opportunities for probe questions from the researcher.

11.6.2.2.2.5   Data Recording and Clustering   Field observations from all the participatory activities, interviews, and photographs were recorded in a common spreadsheet format with a reference to the location of the observation and the participant; his or her story, or the researcher's observation; and any researcher comments or insights on the observation recorded. Figure 11.12 shows a sample record. The data, including photos, then were analyzed using a clustering method in which insights from field observations were grouped into categories of similar issues and themes.

11.6.2.3   Findings

The clustering of field observations identified three high-level categories: Life in the Village, the Healthcare System, and Partner Industries. These were further clustered into a total of 20 categories, each with descriptors, which are shown in Figure 11.13. Brainstorming on each of these 20 themes was done to identify, at a high level, need gaps across all segments and locations. They are:

- Access
  - Qualified consultations
  - Facilities

- Cost
  - Travel
  - Qualified consultations
- Preventive health care
  - Recognition of seriousness of symptoms
  - Information on prevalent diseases

11.6.2.4 Implications and Epilogue

In the second and third phases of the project, the 20 thematic clusters (as shown in Figure 11.13), with all their supporting observations and data, were used to brainstorm new concepts for health-care services and products. Nearly 100 ideas were generated initially. These were reduced to a set of 42 concepts, each of which was developed further in the form of low-cost, often paper prototypes. These were reviewed with stakeholders in the villages and further reduced to a handful of concepts for more formal prototyping and development. A number of these were reported in the Indian media (Indian by Design, 2010; Johnson, 2010).

DATA CLUSTERS

Analyzed data and insights were clustered and organized into three high-level categories: life in the village, the health-care system in rural India, and partner industries.

LIFE IN THE VILLAGE

1. Demographics
   - Migration
   - Financial health (individual)
2. Infrastructure
   - General (power, roads)
   - Medical (equipment, medicine, public/private health, existing supply chain, nongovernmental organizations [NGOs], legislation)
3. Social Structure of the Village
4. Village Economy
   - Financial health (community)
   - Spending patterns
   - Link between health care and employment
5. Environmental Factors
   - Environmental aliments
   - Epidemics
   - Animals
   - Occupational hazards (work environments)
6. Regional Differences
7. Technology

HEALTH-CARE SYSTEM IN RURAL INDIA

8. Health-Care Players and Providers
   - MDs
   - Other health-care workers
   - Entrepreneurs
   - RMPs
   - ANMs
   - Support workers
   - Influencers
   - NGOs
   - Quacks
   - Pharmacists
9. Current Health-Care Options
   - Preventative medicine (includes immunization)
   - Alternative medicine
   - Home cures/devices/self-care
   - First aid
   - Epidemic management
   - NGOs
   - Conventional medicine

FIGURE 11.13   Thematic clusters.   *Continued*

10. Health-Care Chains (System Map)
    - Decision making
    - Medicine delivery
    - Collective decision making (community not individual)
    - Government schemes
    - Influencers
    - Emergency medicine
11. Medical Records
    - Individual medical records
    - Private record keeping
    - Official record keeping (government)
    - Incentives for accurate record keeping
12. Ailments
    - Age related
    - Regional
    - Occupational
    - Seasonal
    - Chronic
    - Congenital
13. Barriers to Seeking Health Care
    - Cost/economic conditions
    - Lack of confidence in doctors
    - Lack of diagnosis
    - Barriers to getting treated
    - Illiteracy
    - Inadequate access to doctors
    - Cultural barriers
14. Rural Beliefs and Expectations
    - Confidentiality for diseases with attached stigma (leprosy, tuberculosis [TB], sexually transmitted diseases [STDs])
    - Omnipresent second opinion
    - Attitude toward illness
    - Aspirations (e.g., preference for female doctors for female patients)
    - Confidence in technology
    - Religious beliefs
    - Allopathic medicine

HEALTH-CARE SYSTEM IN RURAL INDIA

15. Communication
    - Media (e.g., word-of-mouth publicity)
    - Counseling/medical explanations
    - Language
    - Doctor-to-doctor networks
    - Diagnosis and treatment
    - Doctor-to-patient communication
    - Prescriptions

FIGURE 11.13 (*Continued*)   Thematic clusters.   *Continued*

16. Education/Awareness
    - Communication system: educational
    - Training and credentials of medical personnel
    - Awareness of diseases
    - Awareness/understanding of basic health care
    - Illiteracy
    - Hygiene
17. Lifestyle Factors
    - Alcohol consumption
    - Group culture
    - Food/cooking practices
    - Hygiene issues
18. Women's Health
    - Pregnancy
    - Prenatal care/educational system
    - Malnutrition, joint pains
    - Cultural/religious norms
    - ANM
19. Child and Infant Health

PARTNER INDUSTRIES

20. Partner Industries
    - Insurance
    - Health care bundled with other services
    - Telemedicine
    - Microfinance
    - ICT

FIGURE 11.13 (Continued) Thematic clusters.

REFERENCES

Bond, M.H. 1986. *The Psychology of the Chinese People*. New York: Oxford University Press.

Brislin, R.W. 1970. Back-translation for cross-cultural research. *Journal of Cross-Cultural Psychology*, 1: 185.

Collier, J. Jr., and Collier, M. 1999. *Visual Anthropology: Photography as a Research Method*, 6th edition. Albuquerque, NM: University of New Mexico Press.

Farh, J.L., Zhong, C.B., and Organ, D.W. 2004. Organizational citizenship behavior in the People's Republic of China. *Organization Science*, 15: 241–253.

Indian by Design. 2010. Low cost self-diagnosis tool for rural India. May 6. http://indianbydesign.wordpress.com/2010/05/06/product-feature-low-cost-self-diagnosis-tool-for-rural-india/

Johnson, T.A. 2010. Circles of good health. *India Express*. August 8. http://www.indianexpress.com/news/circles-of-good-health/657435/

Kumar, V., and Whitney, P. 2007. Daily life, not markets: Consumer-centered design. *Journal of Business Strategy*, 28: 46–58.

Li, Y.Y., Yang, K.S., and Wen, C.Y. 1985. *Modernization and Chinese Characteristics*, Taipei, Taiwan: Kuei-Kuan.

Plocher, T.A., and Zhao, C. 2002. Photo interview approach to understanding independent living needs of elderly Chinese: A case study. Paper presented at the Fifth Asia-Pacific Conference on Computer-Human Interface, Beijing, November.

Plocher, T.A., Zhao, C., Liang, S.M., Sun, X., and Zhang, K. 2001. Understanding the Chinese user: Attitudes toward automation, work, and life. *Proceedings of the Ninth International Conference on Human-Computer Interaction*, New Orleans, LA, August.

Pulik, L., Bairagi, A., Sardana, A., Trivedi, P., Merchant, S., Gaswami, N., and Ganguly, S. 2007. Creative research as a tool to inform and inspire. *The Exponent*, May 3–7.

Randall, D., Harper, R., and Rouncefield, M. 2010. *Fieldwork for Design*. London: Springer-Verlag.

Rau, P.-L.P., Choong, Y.-Y., and Salvendy, G. 2004. A cross culture study of knowledge representation and interface structure in human computer interface. *International Journal of Industrial Ergonomics*, 34(2): 117–129.

Schaffer, B.S., and Riordan, C.M. 2003. A review of cross-cultural methodologies for organizational research: A best-practices approach. *Organizational Research Methods*, 6(2): 169–215.

SIL International. 2011. *ISO 639-3. Macrolanguage Mappings*. Dallas, TX: SIL International.

Yang, C.F. 1991. 'ziji' of Chinese: Theory and research direction (in Chinese). In *Chinese People and Chinese Psychology—Society and Personality*, ed. C.F. Yang and S.R. Gao, 93–145. Taipei, Taiwan: Yuan-Liou.

Yang, C.F. 2000a. Toward a new conceptualization of Guanxi and Renqing. *Indigenous Psychological Studies in Chinese Societies*, 12: 105–179..

Yang, C.F. 2000b. In the wrong places? Or with the wrong people?: Commentary on "In Search of the Chinese in All the Wrong Places!" *Journal of Psychology in Chinese Societies*, 1: 153–158.

Yang, C.F. 2001a. *How to Study the Chinese: A Collection of Papers on the Indigenous Approach*. Taipei, Taiwan: Yuan-Liou.

Yang, C.F. 2001b. The Chinese conception of the self: Toward a person-making perspective. Paper presented at *Scientific Advances in Indigenous Psychologies: Philosophical, Cultural and Empirical Contributions*. Taipei, Taiwan, October 27–30.

12 Gaining User Acceptance in Specific Cultures

12.1 INTRODUCTION

A newly launched product or service always needs a diffusion time before it is well accepted by the entire target market. One of the important goals of product designers is to shorten this diffusion time and make the product more easily accepted by the target users. Previous studies identified several factors that affect acceptance of technologies. "Perceived usefulness" and "ease of use" are found by Davis (1986) as the main factors for acceptance of a product. Compeau and Higgins (1995) studied the effects of "self-efficacy" on adoption of computers. Taylor and Todd (1995) considered "peer influence" as an important social factor for technology acceptance. Zhao (2008a) makes a compelling argument that peers, organizations, and early adopters have a significant effect on the diffusion and acceptance of technology.

Technology acceptance is especially essential in emerging markets and special user populations such as rural Chinese (Liu et al., 2010; Oreglia et al., 2011; Zhao, 2008a, 2008b) and older adults (Wang et al., 2011). Product managers and designers are always eager to understand users in emerging markets. A problem they face is that, most of the time, technology acceptance of emerging users is different from that of users in well-developed markets. The factors affecting technology acceptance are fundamentally influenced by the special culture of a user group. As a result, the product design logic or marketing focus for developed markets may not be applied well in emerging markets. Also a popular product in one culture may not be accepted well in another. This explains why Chinese people have different preferences than Westerners for some leading Western Internet services, such as Google, Skype, and Groupon.

Besides the common preferences held by a culture, the specific infrastructure situation in an emerging market influences technology acceptance significantly. For example, the revolutionary Apple products iPad and iPhone provide innovative user experiences through wireless Internet or 3G connections. They had great success in the developed world such as in the United States and Europe. However, these products were not accepted so widely in China, except in a few major cities. The main reason was people's low acceptance at that time of wireless Internet and 3G. Although Chinese mobile operators started to provide related services, people still needed time to get familiar with and to build trust in them. The uncertain attitudes in people's minds surely influenced their acceptance for the products that relied on wireless and 3G services.

In emerging markets, literacy is sometimes another serious obstacle to gaining the user's acceptance for a new product. Sukumaran et al. (2009) and Sambasivan et al. (2010) report on case studies in which the constraints of literacy and aversion to technology were absolute barriers to Internet and mobile services in certain communities in rural India. Their solution was to deploy the technology through a local intermediary, such as an agriculture extension agent, who was literate and accepted technology. Local people went to the intermediary with questions that he duly answered after accessing the Internet service. The key was to identify this very early in development of the service so that the design could be tailored to the intermediary and also efficiently support the kinds of questions he expected to receive from his technology-averse constituents. It also established the model for deployment of the mobile Internet service.

When we turn our focus to culture, we should also be aware of the heterogeneity of a culture, even though it is usually understood as a stable and consistent system. In Chapter 2, we introduced Chinese culture's inconsistency between independency and interdependency. This inconsistency can be found between different generations of people, as well as between unevenly developed economic areas, resulting in different attitudes and level of acceptance toward new products. For example, Liu et al. (2010) studied rural Chinese acceptance for mobile entertainment and found different patterns between North China and East China. The northern Chinese culture emphasizes social influence to guide people's adoption decisions, while the eastern Chinese culture emphasizes self-efficacy rather than social factors. As a result, different product design guidelines were required for the two cultures. Therefore, the particular culture of a user group should always be considered when designing a product that caters to the users' acceptance of technology and attitudes.

There are systematic ways to gain user acceptance for a new product, from product planning, user research, product design, to the advertising and sales process. Considering the book's focus, this chapter will pay more attention to the earlier phases of planning, research, and design. Rather than abstract introductions of the ways to understand user acceptance, we prefer to show the process through two case studies. One modeled rural Chinese acceptance for mobile services through interview and explorative survey. The second studied older adult users from three countries and their acceptance for information technology in a laboratory experiment. The main objective of these case studies is to show how to understand user acceptance in a specific culture, and how to design according to those understandings.

12.2 CASE STUDY 1: USING EXPLORATIVE SURVEY TO UNDERSTAND SOCIOECONOMIC INFLUENCES ON RURAL CHINESE USERS' ACCEPTANCE OF MOBILE ENTERTAINMENT

12.2.1 Introduction

The literature on technology acceptance was reviewed by Wang (2010), and the most important acceptance factors were extracted. The researchers then conducted a first round of interviews with rural Chinese users to explore all the important factors that influence users' acceptance for new technologies. Second, they created a survey involving all the factors they found in the first phase and conducted the survey with two large groups of rural Chinese users—one from North China and the other from

East China—to better understand the factors affecting technology acceptance and if the factors' effects were different in different socioeconomic contexts. This case study reviews their findings.

12.2.2 INTERVIEW

In the first phase, interviews were conducted by a researcher who could speak the same dialect as the interviewees. In China, there are dozens of dialects, and most of them are hard to understand by people from other parts. Although the government started to spread the standard Chinese "Putonghua" from the beginning of the 20th century, people still speak their local dialect a lot in daily life, especially in rural areas.

To explore factors influencing user acceptance, a set of open-ended explorative questions were formulated. The questions were mainly about participants' experiences using mobile services including entertainment services, their entertainment life, and the behaviors of surrounding people. The main questions were as follows:

- How do you enjoy your entertainment time in your daily life?
- Do you or people around you use mobile phones? What functions do you/ they use?
- What kind of mobile phone entertainment service have you/they ever used?
- What can urge you to use mobile entertainment services, or why do you want to use them?
- To what extent does the price influence your purchase/service-using decisions?

The qualitative interview data were analyzed to extract the factors influencing users' acceptance of mobile entertainment. The transcripts from phone interviews were printed out, cut into items, and then categorized. At the end, citations reflecting the same factor were pasted together and the factor was named and written above.

Eighteen factors were explored from the interviews. Seventeen among them matched those that the researchers found in the literature. The factors were perceived usefulness, perceived ease of use, perceived complexity, perceived enjoyment/fun, output quality, relative advantage, compatibility, perceived behavioral control, job relevance, voluntariness, innovativeness, technology facilitating conditions, organizational support, visibility, perceived risk, communication facilitating, and perceived novelty.

One new factor that could not be matched to the ones from the literature was defined as "cost." Cost represents the charges of a particular service which could influence users' acceptance. Examples of comments relating to cost were "Is listening to music for free?" and "I hope the service is totally for free."

In addition to these 18 factors, there were 10 factors obtained from the literature that were not expressed by participants during the Phase 1 interviews. This was not surprising because the sample size of the initial interviews was relatively small. It was concluded that the 10 additional factors should not be disregarded. Therefore, all 28 factors were tested and analyzed in the next phase of the survey.

12.2.3   EXPLORATIVE SURVEY

The second phase of the survey was conducted with a questionnaire involving all 28 variables, those found in the interviews, and those from the literature. In order to depict different types of socioeconomic structures in Chinese rural areas, two samples of participants were recruited from two different regions. One is Dezhou in Shandong province, North China. The other is Taizhou in Jiangsu province, East China. Dezhou has a typical traditional extended family society and agriculture-related economic structure. In contrast, Taizhou has an independent small family society and industry-related economic structure, which is very usual in East China. All the participants were either farmers or migrant workers from rural areas in Dezhou or Taizhou. They were recruited either from their houses or their working places. Their experiences with mobile phones and mobile services were also solicited in the questionnaire.

Because most rural users were not familiar with Web-based questionnaires, paper-based questionnaires were provided on a person-to-person basis. In order to keep the validity of each questionnaire, the entire process of answering the questions was assisted by the survey conductor. After filling out the survey, each participant received 20 RMB as a reward.

The two samples were analyzed separately. Exploratory factor analysis was used to find relationships among these factors. Finally, a visualized model was built to describe the results comprehensively.

The results identified the most important factors in the rural Chinese technology acceptance model. Comparing the models of North China and East China, there is an interesting difference in the most important factor of each model, "social influence" for rural people in North China and "self-efficacy" for rural people in East China. The difference was explained by analyzing the socioeconomic structures of each region. The authors asserted that in an agricultural-based economic structure and traditional extended family society, the social responsibility, social insurance, and work collaboration make people interdependent with each other, therefore increasing the importance of social influence when accepting new products. In contrast, in a manufacturing industry economy and small family society, the social responsibility, insurance, and collaboration are more independent among people. People think more about themselves when accepting new things.

12.2.4   DESIGN GUIDELINES

According to the results, Liu et al. (2010) provided several design guidelines for innovative mobile service design aimed at fulfilling user needs revealed by the acceptance model. To address users' acceptance influenced by social factors, the following elements, which help to facilitate positive social impression and social connection, were proposed:

- Involve an interpersonal recommendation function through short message (SM) or Internet.

- Involve an interpersonal invitation function through SM or Internet.
- Let the user know who else is using the product and how many other users there are.
- Recommend related products through social computing, for example, "people who play game A also play game B."
- Design the outward appearance of the product in a way that enhances the users' image to his or her social peers. For example, being associated with a product that looks out of date may convey that same impression of the user by peers.

To encourage a high level of self-efficacy among emerging users, four elements were recommended:

- Make it easy to use.
- Avoid using too many characters on the interface. Use appropriate symbols and icons instead.
- Put a significant number of hotkeys on the main interface.
- Encourage users to feel at ease and confident when introducing a product. For example, reassure users by saying "just press one key" or "don't worry about making mistakes," and so forth.

Additionally, the study addressed the factor of "cost" in the rural Chinese users' acceptance model. Cost is always a central concern of emerging users, especially those from developing countries and from cultures that highly value industriousness. Many cultural variables have effects on people's perception of cost and benefit. For example, long-term– versus short-term–oriented cultures place a higher value on future benefits. Also, they are less inclined to spend money only for instant enjoyment. As another example, people in risk avoidance versus risk allowance cultures have more worries about unexpected or unanticipated costs. In addition, interpersonal trust varies in different cultures. Consumers in low-trust cultures will tend to be more suspicious about the cost. Especially when the price is much lower than expected, low-trust cultures may doubt the quality of the product, or they may suspect the credibility of the producer or the seller. All these can influence people's acceptance and buying decisions for a new product or service. As Chinese is a long-term–oriented, risk avoidance, low-trust culture, cost becomes an essential factor in technology acceptance. In the paper by Liu et al. (2010), the authors proposed three recommendations to ensure affordable costs to the users:

- Explain the possible cost clearly when users first see it.
- Emphasize that the cost is really low. One example could be saying "only 2 RMB per month, equal with the cost of an apple."
- Allow users to get a discount by contributing to the product. For example, if a user invites five or more users for a service, let him or her use it at no cost for 1 month.

12.2.5 IMPLICATIONS

The implication of the study by Liu et al. (2010) for innovative product design is to consider factors affecting people's acceptance and attitudes toward new products. Those factors can vary between different cultures and different socioeconomic contexts. It is important to understand people's key concerns in a specific culture and to address those concerns early in the design process.

Another lesson from the study is that, even though culture is viewed as a stable system, there always is heterogeneity within a given culture. In Chapter 2, we introduced Chinese culture's inconsistency between independence and interdependence. This inconsistency can be found among different generations of people, as well as regions with different socioeconomic characteristics. As we saw in this study, this heterogeneity affects the users' attitudes of acceptance toward new products.

12.3 CASE STUDY 2: A CROSS-CULTURAL STUDY ON OLDER ADULTS' INFORMATION TECHNOLOGY ACCEPTANCE

12.3.1 BACKGROUND

The older adult population has increased rapidly worldwide in recent years. For example, in China, the number of citizens older than 60 years is expected to increase from 12.8% in the year of 2010 to about a third of the national population by the middle of this century (Zhai, 2010). Increasing numbers of older adults are choosing to use computers and computer-related technology in their daily lives. Understanding older adults' technology acceptance behaviors is essential to provide them with the most appropriate products. It would have both economic and social significance to investigate older adults' technology acceptance behaviors, improve information technology design, and encourage more of these older adults to use information technologies.

Research on cultural ergonomics indicates that environmental factors and characteristics that vary by culture can significantly influence individuals' perceptions of and behaviors with information technology. It is essential to investigate the effect of cultural differences on older adults' information technology acceptance, so that information technology can be designed appropriately for older adults from different cultures.

12.3.2 OBJECTIVE

A study by Wang, Rau, and Salvendy (2011) investigated the effect of cultural differences on older adults' information technology acceptance. Davis' Technology Acceptance Model shows that perceived usability and perceived ease of use are two factors that determine individual's technology acceptance. This has been verified by many studies (e.g., Igbaria, Iivari, and Maragahh, 1995; Morris and Venkatesh, 2000). These two factors were investigated in older adults from three countries: the United States, Korea, and China. The researchers (Wang et al., 2011) hypothesized that needs orientation (the use of information technology to access information versus to connect with others), autonomy (low, medium, and high), and facilitating types (family help versus elder-friendly design) can influence the perceived usability and ease of use, and acceptance of technology differently for older adults from the three countries. Guidelines for designers of information technology are provided based on the experimental results.

12.3.3 EXPERIMENT METHODOLOGY

The researchers (Wang et al., 2011) invited older adults, ages 60 to 80 years, from the United States, Korea, and China to a laboratory experiment. The experiments had two phases. First, each participant was introduced to two scenarios and one video. The content of the scenarios and the video differed according to the two experimental conditions of needs orientation, the use of information technology to access information versus to connect with others. For the condition in which information technology is used mainly to access information, the two scenarios were introductions to two wiki websites and a video about a future e-newspaper. For information technology intended to facilitate connecting to others, the two scenarios were introductions to two social networking websites. The video was about a future application that will provide instant communication among family members. Within each pair of scenarios (either the wiki websites or the social networking websites), one scenario was described as "older adult friendly." The other was described as being used simultaneously by their children and grandchildren from whom older adults could get help. After being exposed to the scenario or video, participants were required to answer a questionnaire to rate their interest, intention, and self-estimated frequency of use.

For the second part of the experiment, the participants were asked to do searching tasks on three online map websites. The number of functional items was different on the interface of each of the three online map websites to represent three levels of autonomy. Participants rated their interest and intention to use each website according to their first impression.

12.3.4 RESULTS

The findings from the experiment were as follows:

1. Compared with their Chinese and Korean peers, older adults in the United States scored significantly higher on their interest in and intention to use information technology to access information than to use information technology to connect with others.
2. Compared with their U.S. peers, Chinese and Korean participants had significantly higher scores for interest, intention, and self-estimated frequency of use for information technology to connect with others.
3. Compared with their Chinese and Korean peers, older adults in the United States had significantly more intention to use information technology with a higher degree of autonomy than those with a lower degree of autonomy.
4. U.S. participants were significantly more interested in information technology especially designed for older adults than information technology with accessible family help.

5. Compared with their U.S. peers, Chinese and Korean participants had significantly higher interest in using information technology with accessible family help.

12.4 CONCLUSION

It is generally considered that information technology needs to be simple and concise when the users are older adults. The research by Wang et al. (2011) showed that this view is not completely accurate. Information technology (IT) design should also take into account cultural differences. Cultural differences were found in the effects of needs orientation, autonomy, and facilitation on information technology acceptance. In individualistic cultures like the United States, older adults more readily accept IT products that provide them with information, allow high levels of autonomy, and are designed especially for their age group. On the contrary, older adults from collectivistic cultures like China and Korea are more likely to use information technology that connects them with others, provides an uncomplicated interface with a medium level of autonomy, and facilitates obtaining help from others. These results suggest that older adults' cultural backgrounds are important in the design of information technology.

12.4.1 DESIGN IMPLICATIONS

Seven design guidelines were generated from the research:

1. For older adults in individualistic cultures, develop more products to access information, as well as functions facilitating the collection, searching, storing, distribution, and classification of information, such as memory cues, history records, operation hints, and feedbacks that would help older adults access and process information.
2. For older adults in collectivistic cultures, develop information technology with functions to connect with others, as well as functions that facilitate communication, such as various forms of messaging (text messages or audio and video clips), voice input, and automatic mobile phone updates from websites.
3. Older adults in the United States should be provided information technology with a higher degree of autonomy, such as options for advanced searching or customizable interfaces.
4. For Chinese and Korean older adults, information technology with an uncomplicated interface should be designed.
5. Older adults who have a higher education background should be provided with information technology with a higher degree of autonomy.
6. For U.S. older adults, user-friendly IT products should be specifically designed for their age group.
7. For Chinese and Korean older adults, functions should be provided to facilitate getting help from others without interruption of ongoing operations, for example by voice messaging.

REFERENCES

Compeau, D.R., and Higgins, C.A. 1995. Computer self-efficacy: Development of a measure and initial test. *MIS Quarterly*, 19: 189–211.

Davis, F.D. 1986. A technology acceptance model for empirically testing new end-user information systems: Theory and results. Doctoral dissertation, MIT Sloan School of Management, Cambridge, MA.

Igbaria, M., Iivari, J., and Maragahh, H. 1995. Why do individuals use computer technology? A Finnish case study. *Information and Management*, 29: 227–238.

Liu, J., Liu, Y., Rau, P.L.P., Li, H., Wang, X., and Li, D.J. 2010. How socio-economic structure influences rural users' acceptance of mobile entertainment. In *Proceedings of the 28th International Conference on Human Factors in Computing Systems (CHI2010)*, 2203–2212. Atlanta, GA: ACM Press.

Morris, M.G., and Venkatesh, V. 2000. Age differences in technology adoption decisions: Implications for a changing work force. *Personnel Psychology*, 53: 375–403.

Oreglia, E., Liu, Y., and Zhao, W. 2011. Designing for emerging rural users: Experiences from China. In *Proceedings of the 29th International Conference on Human Factors in Computing Systems (CHI2011)*, 1433–1436. Atlanta, GA: ACM Press.

Sambasivan, N., Cutrell, E., Toyama, K., and Nardi, B. 2010. Intermediated technology use in developing communities. In *Proceedings of the 28th International Conference on Human Factors in Computing Systems (CHI2010)*, 2583–2692. Atlanta, GA: ACM Press.

Sukumaran, A., Ramlal, S., Ophir, E., Kumar, V.R., Mishra, G., Evers, V., Balaji, V., and Nass, C. 2009. Intermediated technology interaction in rural contexts. In *Proceedings of the 27th International Conference Extended Abstracts on Human Factors in Computing Systems (CHI2009)*, 3817–3822. Atlanta, GA: ACM Press.

Taylor, S., and Todd, P.A. 1995. Understanding information technology usage: A test of competing models. *Information Systems Research*, 6(2): 144–176.

Wang, L. 2010. Variables contributing to older adults' acceptance of information technology in China, Korea, and USA. Doctoral dissertation, Tsinghua University, July.

Wang, L., Rau, P.L.P., and Salvendy, G. 2011. A cross-culture study on older adults' information technology acceptance. *International Journal of Mobile Communications*, 5(9): 421–440.

Zhai, J. 2010. A new method to deal with the problem of national population aging. *Policy Research*, 14–15 (In Chinese).

Zhao, J. 2008a. ICT4D: Internet adoption and usage among rural users in China. *Knowledge, Technology, and Policy*, 21: 9–18 (DOI 10.1007/s12120-008-9041-0).

Zhao, J. 2008b. Integrating the Internet into farming activities. *Science, Technology, and Society* 12: 2, 325–344.

13 International Usability Evaluation

13.1 MITIGATING CULTURAL BIAS IN MODERATOR–USER INTERACTIONS

Usability testing inherently involves social interaction between a test moderator and a test user. Social and cultural norms affect this interaction in a similar manner to the way that they affect other social interactions. There is growing literature on how Easterners and Westerners behave in usability tests. A number of best practices can be described that help to mitigate cultural bias in usability tests resulting from social interaction effects.

First, it is a good practice to use moderators or interviewers from the same culture as the test users. Vatrapu and Pérez-Quiñones (2006) studied how test users from different cultures behaved in a structured interview setting. Their task was to comment on a website. Indian test users who worked with an Indian moderator found more usability problems and made more suggestions than those who worked with an Anglo-American moderator. Further, the comments they made to the Anglo-American moderator tended to be more positive than negative. With an Indian moderator conducting the session, Indians were more inclined to discuss culture-related problems with the website and were more detailed and candid. With an Anglo-American, they kept their comments quite general. Yammiyavar, Clemmenson, and Kumar (2008) found that when subjects were paired with test moderators from the same culture, they used more head and hand gestures to communicate than if the moderator was from a different culture, providing a richer source of nonverbal data to analyze. Sun and Shi (2007) studied how using one's primary versus secondary language (English versus Chinese in her study) in a think-aloud test affected the process of the test. Chinese moderators speaking Chinese to Chinese test users gave a more complete introduction to the product being tested and, during the test, encouraged users more frequently than did Chinese moderators who spoke English during the test.

A second good practice is to avoid pairing moderators and test users who differ in their perceived status or authority. Particularly in cultures with high power distance, such as China and Malaysia, the behaviors of both the test user and the test moderator are affected by perceived differences in status or authority. Test users in high power distance cultures do not challenge or question the test moderator because of the perception that the moderator is a person of authority (Burmeister, 2000). Yeo (1998) relates the story of a test user in Singapore who broke down and cried from frustration during the test. The posttest interview revealed that the test user, though frustrated to tears by the task, still did not consider it acceptable to criticize the designer openly and cause him to lose face. Evers (2002) evaluated cultural differences between four culturally different user groups (England, North America, the Netherlands, and Japan) in understanding a virtual campus website. She found that Japanese test users who were secondary school students felt uncomfortable speaking their thoughts out loud. They also reported feeling insecure because they could not confer with their peers to reach a common group opinion.

The effect of perceived authority can go both ways. In a study of think-aloud tests, Sun and Shi (2007) found that the moderator's behavior is also affected by differences in level of perceived authority. They observed that when the evaluator's academic title or rank was higher than that of the users, the evaluator more frequently asked the test user what he or she was thinking during the test.

The third guideline is to train test moderators to combat the "conversational indirectness" of Asian users. When evaluating user interfaces in a test situation, Easterners tend to be neutral and indirect. For example, Herman (1996) studied cultural effects on the reliability of objective and subjective usability evaluation. The results of objective and subjective evaluation correlated poorly in Herman's study. The Asian participants were less vocal, very polite, and not inclined to express negative comments in front of observers. The results of the subjective evaluation tended toward the positive despite clear indications of poor user performance. Herman's solution was to invite test participants to work in pairs to evaluate the interface and make the usability test more of a peer discussion session.

In a study conducted with rural villagers in Ghana, Gorman et al. (2011) found that allowing users to evaluate a user interface in pairs or small groups was more successful than individual testing because it more closely mimicked the way in which villagers normally used such devices. Placing members from the same family or close friends together in a group generally was successful, as long as one member of the group was not traditionally marginalized by another member of the group, for example, a wife by her husband.

Shi (2008) conducted observations of usability tests in China, India, and Denmark, and like Herman, also noted that Chinese users often kept silent and did not speak out actively, particularly in formative evaluations. Shi (2008) and Clemmensen, Hertzum et al. (2009) explained this observation in terms of Nisbett's (2003) cultural theory of Eastern and Western cognition. According to that theory, Chinese people tend to have a holistic process for thinking as opposed to the more analytic style of Westerners. Holistic thought is not as readily verbalized as analytic thought. So it may be speculated that, in a think-aloud situation, Chinese users are thinking about the user interface in holistic terms which they simply have more difficulty expressing in spoken words.

Shi recommended that moderators receive special training to conduct think-aloud tests with Chinese users. Moderators should be trained to use reminders and questions, such as "digging deeper probes," to get the test users to talk. For example, moderators in Shi's study reported that if they knew that users were looking for some object or feature on the screen, they would ask, "what are you looking for?" The user would tell them immediately about what he or she was looking for. This method of asking related questions to encourage speaking aloud was found to be more natural than just asking people to "keep talking." Shi (2009) also points out that test users from Eastern cultures often pause and think in between verbalizations. She suggests that moderators should be trained to recognize this tendency and to exercise patience with test users (Shi, 2009).

Chavan (2005) devised a method for helping test users express their emotions toward the product more privately, resulting in freer expression during the test. She borrowed the notion of "rasas" from Indian performing arts. Rasas are classes of emotions on which most traditional Indian performing arts are based. There are nine rasas or emotions, ranging from love to anger to astonishment. Chavan operationalized these rasas in the form of "emotion tickets," similar to movie tickets. Each of the nine emotion tickets was presented with highly associative images and dialogue from Bollywood movies. During the usability evaluation session, test users would use the tickets to express the emotion they felt when they used some particular feature of the user interface. Then they would make a note about why they felt that way. The method capitalized on the popularity in India of watching Bollywood movies to motivate test users, while at the same time creating a safe situation for them to express their feelings about the product to the moderator.

So does the same cultural tendency toward reporting problems also affect test moderators from Eastern cultures? Shi (2009) found no significant differences in the set of usability problems found by Chinese and Danish evaluators. However, their ratings of the severity of the usability problems were significantly different. Chinese evaluators rated problems less severely than Danish evaluators and often rated problems in the middle of the five-point severity scale. One solution to this problem of neutrality is simply to remove the opportunity for mid-scale scoring of severity by using a four-point, rather than five-point scale (Shi, 2009).

13.1.1 TEST USER AND MODERATOR RECRUITING

The need to recruit test users with similar backgrounds but in different target locales can make international usability evaluation logistically difficult. A number of options for conducting international usability evaluations can be considered. Test moderators can travel to the target locales and personally conduct the tests in-country. Local assistance may be enlisted from a local usability consultant or staff from your company's local branch office. In some cases, tests can be run remotely (Nielsen, 1990, 2003). Many researchers (Choong and Salvendy, 1998, 1999; Dong and Salvendy, 1999; Evers, 2002; Fang and Rau, 2003; Fukuoka, Kojima, and Spyridakis, 1999; Prabhu and Harel, 1999) chose to recruit test users in two or more countries by actually traveling to the countries.

Clemmensen, Shi et al. (2007) address the problem of "hidden user groups." These are groups of people who represent significantly different target user segments within the same culture. They suggest that test planners attempt to balance out potential hidden user groups within user segments. For example, test users who are accustomed to foreigners and adapt quickly to international test conditions should be balanced by users who are not accustomed to foreigners. Traditional and culturally sensitive users, such as those one might find in rural areas, should be balanced in the pool of test users by more modern, urbanized people who are less influenced by a country's local culture. To avoid missing critical usability problems during the test, they recommend that test moderators be chosen carefully so as to be compatible with the identified hidden user groups.

13.1.1.1 Language

13.1.1.1.1 Verbal Language and the Use of Interpreters and Translators

Language is one of the most significant factors in international usability evaluation. Without translated content, the target audience may be limited to Web users with a certain education and social background. Nielsen (2003) suggests that the top two concerns for international usability testing are displaying information in the user's native language, character set, and notations, and translating the user interface and its documentation into the user's native language. Rau and Liang (2003b) conducted a card-sorting test with Chinese users in Taiwan for an international website. Even though all the information items were translated into Chinese, a dictionary and instructions were available if participants had any question about the meaning of these items. Also, any language considered offensive in the test users' cultural background should be avoided. The testing materials and procedure should accommodate the way business is conducted and the way people communicate in the target locale (Nielsen, 2003). If the test moderators are not able to speak the test user's native language well, then interpreters are essential. Nielsen (2003) suggests meeting interpreters beforehand and reminding them that they should not help users during the test.

13.1.1.1.2 Usability Testing with Oral Users

Most of our usability testing methodology is based on experience with literate test users (e.g., those who have both written and oral language skills). Sherwani and colleagues (2009) described the challenges presented to traditional HCI design methodology by target users who have an oral, rather than written and oral, language tradition.

The experiences of Gorman et al. (2011) in evaluating an audio recorder-player with rural villagers in Ghana confirmed the challenges described by Sherwani. They applied numerous modifications to traditional usability evaluation methodology to address these challenges. First, people from an oral tradition think, learn, and process information differently than users of written language. Gorman et al. (2011) concluded that it would be difficult for a nonindigenous test moderator to assimilate such cognitive differences and then facilitate the test and interpret the results appropriately. Their solution was to train local people to be test moderators. Second, users from an oral tradition tend to struggle with assimilating information from abstract presentations of instructions and scenarios. The solution described by Gorman et al. (2011) was to explain the device to the test users within a context they were all familiar with, in this case, farming. The moderator led a discussion with test users about the complexities of farming and the lack of agricultural extension support. The value of a device that could present useful farming information then was well understood by everyone and they wanted to learn more. Third, oral communication relies on repetition to ensure that a message is understood. The methods that Gorman et al. (2011) used to train subjects relied heavily on repeated practice, as did the test moderator's interactions with the participants. If the participant struggled with a step or action and the moderator was forced to give clues, it was done in a manner similar to the initial training, capitalizing on the test user's comfort with and need for communication redundancy.

13.1.1.1.3 Nonverbal Language

Test users occasionally use nonverbal language such as hand and head gestures as a substitute for verbal language or to elaborate on or supplement verbal remarks. Hand and head gestures also may communicate the test user's level of comfort with the test situation and his or her readiness to communicate with the moderator. Observing and analyzing gestures during a test can provide a rich source of data that add context to the verbal data being recorded.

Yammiyavar et al. (2008) questioned whether test users from different cultures exhibited similar or different patterns of nonverbal communication including the type, frequency, and usage of gestures. They analyzed gestures documented in video recordings of think-aloud tests. Test users were from Denmark, India, and China. Gestures were grouped into four types, using the classification of Ekman and Friesen (1969). They included emblems, illustrators, adapters, and regulators. Emblems are gestures, such as nods of the head for "yes" or a V sign for victory, that replace words. Emblems tend to be culture specific. Illustrators are actions such as banging on the table or sketching shapes in the air. They help test users verbalize their thoughts. Yammiyavar et al. discovered that these illustrator gestures can be quite important markers for usability problems because they frequently precede the verbalization of a usability problem to the moderator. Adapter gestures are actions of the body that convey feelings of pressure or discomfort. They might take the form of cracking one's knuckles, tapping one's feet, stroking one's hair or chin while in deep contemplation, or "squirming" in one's seat. Adapters indicate the test user's comfort level with the test situation. Finally, test users control the flow of conversation with the moderator by using regulator behaviors, such as nodding the head up and down to indicate agreement.

Yammiyavar et al. (2008) found that the frequency of using gestures during the test was not significantly different across the three cultural groups—Danes, Indians, and Chinese. This contradicts the popular belief that Indians, for instance, use more gestures to communicate than other cultural groups. Regulator gestures were used similarly across the cultures studied, but with some tendency for Chinese to use them the least. In contrast, there were significant differences across the cultural groups in the specific emblem gestures used to replace words and in adapter behaviors. Certain illustrators appeared to be culture specific, as well. The researchers concluded that there is a need to benchmark gestures used in these different cultures and their meanings, and then provide those to usability test moderators to guide observations and better understand what they are observing.

During a usability test, moderators often look for a certain facial expression associated with the emotion of "surprise." The common belief is that this expression indicates that a usability problem was detected by the test user. Clemmensen, Hertzum et al. (2009) questioned the validity of this practice in cross-cultural usability tests. From Nisbett's (2003) theory of cultural cognition, they hypothesize that Easterners actually will experience less surprise than Westerners when presented with inconsistencies in user interfaces. With their logical, analytic orientation, Westerners tend to focus on fewer causes of observed events. Easterners, with their holistic orientation, tend to consider more causal factors as well as the context of the event. Clemmensen, Hertzum et al. (2009) point out that this makes it easier for Easterners to identify a rationale for the occurrence of an event during the test, resulting in less surprise.

13.1.2   Instructions, Tasks, and Scenarios

The amount of contextual information contained in instructions can vary significantly. At the one extreme, instructions might be strictly focused on the task to be performed with the application being tested. No explanation is given about the purpose of the application, when you might use it, or why. At the other extreme, the explanation of the task might be embedded in a rich context provided by a real-life scenario. Clemmensen et al. (2009) suggest that Westerners, with their tendency to focus on the central elements presented such as the details of the task, will be able to obtain that regardless of the amount of context provided in the test instructions. Easterners, however, with their holistic style of thinking, will prefer to have the task explanation embedded in the context of a real-life scenario. The recommendation from Clemmensen et al. (2009) is that in cross-cultural usability testing, the test planner should consider adapting the instructions to the culture of the test users. Planners might want to have different versions of the test protocol prepared that include different types and amounts of background information.

13.1.3   Templates for Cross-Cultural Usability Evaluation

Clemmensen (2011) was intrigued by the possibility that a significant portion of the "inherent subjectivity" of usability tests proposed by Vermeeren (2008) was not due to random or arbitrary factors, but rather is actually related to variations in the parts and contexts of usability tests in different countries. He went on to investigate how usability tests are administered by usability consultancy or vendor companies in three different countries, Denmark, China, and India. From the interviews, a cross-cultural taxonomy of a think-aloud usability test was developed. As shown in Figure 13.1, the taxonomy consists of a matrix that relates four parts of a usability test to eight usability test contexts for a total of 32 possible aspects of a test. From in-depth interviews with test administrators, Clemmensen (2011) determined that 21 of these aspects were considered important aspects of the usability test in all three countries. These 21 common important test aspects are highlighted graphically in black in Figure 13.1. These 21 test aspects describe a basic cross-cultural template for usability tests. Faced with planning and conducting a cross-cultural usability test, a practitioner can use this template as a checklist for factors that must be considered in the planning. For example, the practitioner would have to consider the "interaction of instructions and tasks" with the "user's age and personality" and "the overall user-evaluator relationship" to the "design of test/formative evaluation."

In addition to these 21 common aspects of a usability test, Clemmensen (2011) discovered aspects of usability testing that were unique to practitioners in each country.

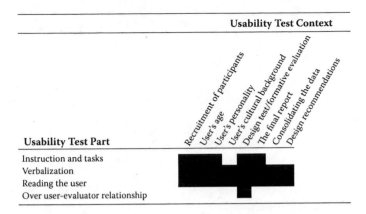

FIGURE 13.1 Template for a cross-cultural usability test. (From Clemmensen, T. 2011. Templates for cross-cultural and culturally specific usability testing: Results from field studies and ethnographic interviewing in three countries. *International Journal of Human-Computer Interaction*, 27: 7, 634–669 (DOI: 10.1080/10447318.2011.555303, First posted on March 4 [iFirst]). With permission.)

That is, practitioners in each country tended to include their own additional test parts and test contexts to the way they planned and executed a usability test. In other words, they adapted to the expectations and ecosystem involved in practicing the art and science of usability testing in their national culture. Three test archetypes were identified—"Evaluator-centered" in China, "User-centered" in India, and "Client-centered" in Denmark—referring to the general orientation of these additional test aspects in each country.

The Evaluator-centered archetype in China placed great emphasis on test methodology and the evaluator's role in selecting a methodology that was tailored carefully to user contexts and to the specific type of application to be tested. Figure 13.2 shows the template for the Evaluator-centered archetype.

In India, the "User-centered" archetype placed a strong emphasis on aspects of the test protocol related to greeting and informing the test user. It also focused attention on the relationship between the test user and evaluator, emphasizing the need to adapt the test protocol and select evaluators such that a test environment is created in which the user feels comfortable. A template for the "User-oriented" test is illustrated in Figure 13.3.

The "Client-centered" archetype adds an important test part related to evaluator experience with the client. Moderator experience cuts across the majority of the test contexts, including familiarity with the client's business goals and measures of success. The "Client-centered" template is shown in Figure 13.4.

Practitioners can use these templates as a reference. The templates can provide some guidance about additional test parts and test contexts that the practitioner might wish to consider in addition to those described in the basic cross-cultural test template. Note that this is a very new and novel approach to systematizing cross-cultural usability testing. It remains for future studies to investigate to what extent practitioners can work with multiple templates in multiple country usability testing.

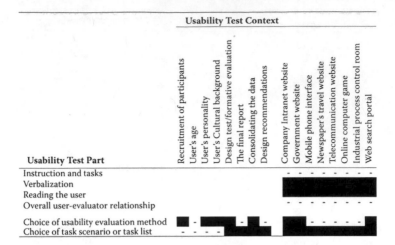

FIGURE 13.2 Evaluator-centered usability test archetype for a test in Beijing. (From Clemmensen, T. 2011. Templates for cross-cultural and culturally specific usability testing: Results from field studies and ethnographic interviewing in three countries. *International Journal of Human-Computer Interaction*, 27: 7, 634–669 (DOI: 10.1080/10447318.2011.555303, First posted on March 4 [iFirst]). With permission.)

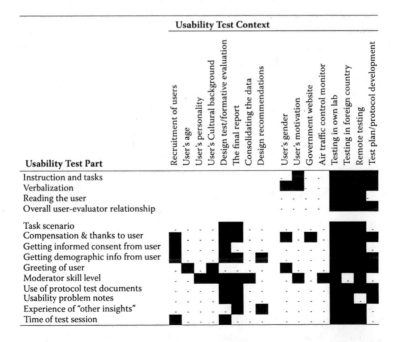

FIGURE 13.3 User-centered usability test archetype for a test in Mumbai. (From Clemmensen, T. 2011. Templates for cross-cultural and culturally specific usability testing: Results from field studies and ethnographic interviewing in three countries. *International Journal of Human-Computer Interaction*, 27: 7, 634–669 (DOI: 10.1080/10447318.2011.555303, First posted on March 4 [iFirst]). With permission.)

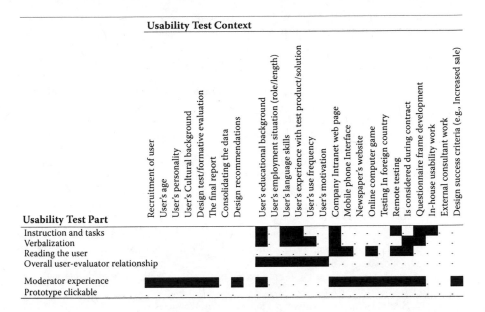

FIGURE 13.4 Client-centered usability test archetype for a test in Denmark. (From Clemmensen, T. 2011. Templates for cross-cultural and culturally specific usability testing: Results from field studies and ethnographic interviewing in three countries. *International Journal of Human-Computer Interaction*, 27: 7, 634–669 (DOI: 10.1080/10447318.2011.555303, First posted on March 4 [iFirst]). With permission.)

REFERENCES

Burmeister, O.K. 2000. Usability testing: Revisiting informed consent procedures for testing Internet sites. Paper for the Second Australian Institute of Computer Ethics Conference, Canberra, Australia, December.

Chavan, A.L. 2005. Another culture, another method. In *Proceedings of the 11th International Conference on Human-Computer Interaction* [CD-ROM]. Hillsdale, NJ: Erlbaum.

Choong, Y.Y., and Salvendy, G. 1998. Design of icons for use by Chinese in Mainland China. *Interacting with Computers,* 9: 417–430.

Choong, Y.Y., and Salvendy, G. 1999. Implications for design of computer interfaces for Chinese users in Mainland China. *International Journal of Human-Computer Interaction*, 11: 29–46.

Clemmensen, T. 2011. Templates for cross-cultural and culturally specific usability testing: Results from field studies and ethnographic interviewing in three countries. *International Journal of Human-Computer Interaction*, 27: 7, 634–669 (DOI: 10.1080/10447318.2011.555303, First posted on March 4 [iFirst]).

Clemmensen, T., Hertzum, M., Hornbæk, K., Shi, Q., and Yammiyavar, P. 2009. Cultural cognition in usability evaluation. *Interacting with Computers*, 21: 212–220.

Clemmensen, T., Shi, Q., Kumar, J., Li, H., Sun, X.H., and Yammiyavar, P. 2007. Cultural Usability Tests—How usability tests are not the same all over the world. In *Usability and Internationalization*, Part I, HCII 2007, ed. N. Aykin, LNCS 4559, 281–290. Berlin: Springer-Verlag.

Dong, J., and Salvendy, G. 1999. Designing menus for the Chinese population horizontal or vertical? *Behaviour and Information Technology*, 18(6): 467–471.

Ekman, P., and Friesen, W.V. 1969. The repertoire of nonverbal behavior: Categories, origins, usage, and coding. *Semiotica*, 1: 49–98.

Evers, V. 2002. Cross-cultural applicability of user evaluation methods: A case study amongst Japanese, North-American, English and Dutch users. Paper for CHI 2002, Minneapolis, MN, April 20–25.

Fang, X., and Rau, P.L.P. 2003. Culture differences in design of portal sites. *Ergonomics*, 46: 242–254.

Fukuoka, W., Kojima, Y., and Spyridakis, J.H. 1999. Illustrations in user manuals: Preference and effectiveness with Japanese and American readers. *Technical Communication*, 2nd Quarter.

Gorman, T., Rose, E., Yaaqoubi, J., Bayor, A., and Kolko, B. 2011. Adapting usability testing for oral, rural users. *CHI 2011*, Vancouver, BC, Canada, May 7–12. ACM 978-1-4503-0267-8/11/05.

Herman, L. 1996. Toward effective usability evaluation in Asia: Cross cultural differences. *INTERACT '96/OZCHI*, 1–7, 135.

Nielsen, J. 1990. Usability testing of international interfaces. In *Advances in Human Factors/ Ergonomics*, *13, Designing User Interfaces for International* Use, ed. J. Nielson, 39–44. New York: Elsevier Science.

Nielsen, J. 2003. International Usability Testing. http://www.useit.com/papers/international_ usetest.html

Nisbett, R.E. 2003. *The Geography of Thought: Why We Think the Way We Do*. New York: Free Press.

Rau, P.-L.P., and Liang, S.-F.M. 2003b. Internationalization and localization: Evaluating and testing a Web site for Asian users. *Ergonomics*, 46: 255–270.

Sherwani, J., Sherwani, J., Ali, N., Penstein Rosé, C., and Rosenfeld, R. 2009. Orality-grounded HCID: Understanding the oral user. *Information Technologies and International Development*, 5: 37–49.

Shi, Q. 2008. A field study of the relationship and communication between Chinese evaluators and users in thinking aloud usability tests. *NordiCHI 2008*, Lund, Sweden, October 18–22.

Shi, Q. 2009. An empirical study of thinking aloud usability testing from a cultural perspective. PhD thesis, LIMAC PhD School, Programme in Informatics, Copenhagen Business School (CBS), Copenhagen, Denmark (November version).

Shi, Q., and Clemmensen, T. 2008. Communication patterns and usability problem finding in cross-cultural thinking aloud usability testing. CHI '08 extended abstracts on human factors in computing systems, ACM, Florence, Italy, 2811–2816.

Sun, X., and Shi, Q. 2007. Language issues in cross cultural usability testing: A pilot study in China. *Lecture Notes in Computer Science*, 4560: 274–284.

Vatrapu, R., and Pérez-Quiñones, M. 2006. Culture and Usability Evaluation: The Effects of Culture in Structured Interviews. *Journal of Usability Studies*, 1: 156–170.

Vermeeren, A.P.O.S., Attema, J., Akar, E., de Ridder, H., von Doorn, A.J., Erbug, C., and Maguire, M.C. 2008. Usability problem reports for comparative studies: Consistency and inspectability. *Human–Computer Interaction*, 23: 329–380.

Yammiyavar, P., Clemmensen, T., and Kumar, J. 2008. Influence of cultural background on non-verbal communication in a usability testing situation. Special issue on cultural aspects of interaction design. *International Journal of Design*, 2: 31–40.

Yeo, A. 1998. Cultural effects in usability assessment, doctoral consortium. *Proceedings of the conference on CHI '98 summary: Human factors in computing systems,* April, 71–75.

14 Heuristic and Guidelines-Based Evaluation in Cross-Cultural Design

14.1 INTRODUCTION TO HEURISTIC EVALUATION

Usability evaluation with local users requires the commitment of a significant amount of time and money. Further, as we discussed in Chapter 13, it is a not a trivial undertaking to organize a cross-cultural usability test, given the requirement for appropriate users, appropriate evaluators, and a test approach that is well adapted to the local cultures of interest. Cross-cultural usability testing is absolutely essential to ensure that the product has high usability in the target cultures. However, usually there are practical limits to how much formal usability testing can be done during product development.

Heuristic evaluations of a user interface design against established principles and guidelines provide an inexpensive method for periodically checking the goodness of a user interface design from initial concept prototypes in Phase 2 to final design validation in Phase 5 (Nielsen, 1993, 1994). Heuristic evaluations can be done fairly quickly, do not require a test environment to be set up or subjects to be recruited, and may actually discover issues in the design that might not be noted by users. It also fits extremely well with a philosophy of iterative product design. If performed after each major iteration of the user interface design during the development process, it can provide a rigorous method for tracking the level of usability attained by that version of the product prototype. It identifies where the usability problems lie in the product and highlights them so that they can be addressed by the development team. Heuristic evaluation results, particularly if presented in a quantitative or semiquantitative manner, are always a welcome contribution to management and phase gate reviews.

That said, human factors engineers should use heuristic evaluation with an understanding of its limitations. First, the human factors experts conducting the heuristic evaluation will not necessarily be domain experts. This is generally not an issue for consumer products. Consumer products reside in domains that we all are familiar with, such as personal communications, social networking, Web commerce, and so forth. However, some products will reside in highly technical domains, such as industrial control or aviation applications, and require a certain amount of domain expertise to evaluate the user interface design. Second, one human factors expert might not discover all the usability problems in the design. Generally assume that at least two evaluators will be needed to discover a significant number of the problems by means of heuristic evaluation.

14.2   A GENERAL-PURPOSE HEURISTIC
EVALUATION SCORECARD

Burns and Hajdukiewicz (2004) proposed a hierarchical classification of usability heuristics that is based on seven areas of usability: content, functionality, information architecture and navigation, system responsiveness, user freedom and control, user guidance and support, and visual design. The heuristic criteria proposed under each of these seven usability areas are shown in Table 14.1. Dharwada and Tharanathan (2011) further developed this framework into a heuristic evaluation scorecard. This scorecard produces not only a list of usability problems or defects by area and heuristic, but also a quantified score for each heuristic. Comparing the scores across all the heuristics clearly highlights the usability strengths and weaknesses of the product. This provides focus for the development team.

14.2.1   Using the Scorecard

Table 14.2 illustrates a single finding from a product evaluation using the scorecard. The steps in conducting an evaluation with the scorecard are as follows:

1. Obtain the most current prototype of the product.
2. Define the scenarios of use that will be used to exercise the prototype in the evaluation.
3. Exercise the prototype using the scenarios; identify usability problems using the table of usability areas and heuristics as a guide.
4. Describe each identified usability problem and place it under one of the seven areas of usability.
5. Identify the usability heuristic guideline that the problem violates and place it under that heuristic.
6. Rate each problem for importance using three risk criteria: severity, frequency of occurrence, and detectability. A value of 1 (low risk), 3 (medium risk), or 9 (high risk) is assigned to each of the three rating criteria.
7. Using the computational process shown below and illustrated in Tables 14.3 and 14.4, combine the risk ratings into a single risk score for each heuristic and, overall, for each of the seven usability areas.
8. Describe a solution for each problem, and identify an individual on the development team to address the problem.

14.2.2   Scoring and Reporting the Findings

14.2.2.1   Scoring

The heuristic evaluation findings can be scored or used in numerous ways. The reader is encouraged to adopt a scoring scheme that makes sense to his or her need for quantitative reporting on the development project.

In simplest form, the findings can be presented as the number of usability problems discovered within each heuristic and usability area. Areas and heuristics with more problems will be obvious and suggest design features that need attention.

TABLE 14.1

Usability Areas and Heuristic Guidelines

	Definitions	
Usability Area	**Usability Heuristics**	**Description**
Content	Relevant and precise content	The system provides content of information that is relevant to user's context and presents in appropriate areas.
Functionality	Unambiguity/ appropriateness/ availability of required functionality	The system provides required functionality to the user. Specific functionality of the system is appropriate and useful, and there is no ambiguity to the user.
Information architecture	Menu structures and hierarchy	Structuring of information into menus matches user's mental model. Menu items, structure, and hierarchy are intuitive and easy to learn.
	Organization/home screen layout	Home screen provides the user with a clear purpose of the system and defines the organization of the application. User actions on the system are made clear, and it gives direct access to key content/tasks (search box, etc.).
Navigation	Avoid deep navigation	Key content and functions are visible in first screenshot and do not require scrolling or deeper levels of navigation.
	Clear signs, cues, and aids for navigation and orientation	The system provides clearly visible/understandable elements that help users to orient within the system navigation and context. The system provides navigation aids to efficiently navigate through long lists, trace back the path taken, or return to the primary context within a task.
System responsiveness	Highly responsive	The system responds very quickly, user does not have to wait too long for the system to provide a response.
User control and freedom	Error prevention, recovery, and control	Even better than good error messages is designing for prevention of a problem from occurring in the first place. Either eliminate error-prone conditions or check for them and present users with a confirmation option before they commit to the action. Users often choose system functions by mistake and will need a clearly marked "emergency exit" to leave the unwanted state without having to go through an extended dialogue (Recovery).
	Flexibility, control, and efficiency of use	The system provides accelerators—unseen by the novice user—that may often speed up the interaction for the expert user such that the system can cater to both inexperienced and experienced users. Allow users to tailor frequent actions. Provide additional quick ways of supporting user workflow practices.
User guidance and support	Consistency	The system follows standard user conventions across the application. Users should not have to wonder whether different words, situations, controls, or actions mean the same thing. The system behavior corresponds to user expectations.

Continued

TABLE 14.1 (*Continued*)
Usability Areas and Heuristic Guidelines

	Definitions	
Usability Area	**Usability Heuristics**	**Description**
	Compatibility	Match between real world and system status. There is compatibility with metaphors chosen for presentation.
	Informative feedback and status indicators	The system always keeps users informed about what is going on through appropriate feedback and when needed provides a course of action within a reasonable time.
	Recognition rather than recall	Minimize the users' memory load by making objects, actions, and options visible. The user should not have to remember information from one part of the dialogue to another. Instructions for use of the system should be visible or easily retrievable whenever appropriate.
	Terminology: informative titles, labels, prompts, and messages	The system speaks the users' language, with words, phrases, and concepts familiar to the user, rather than system-oriented terms. It follows real-world conventions, making information appear in a natural and logical order.
	Tool tips, help, and documentation	The system provides tool tips where necessary. Even though it is better if the system can be used without documentation, it may be necessary to provide help and documentation. Any such information should be easy to search, focused on the user's task, list concrete steps to be carried out, and not be too large.
	Work-flow support	System work-flow is compatible to user mental model of the workflow and guides the user along the workflow.
Visual design	Aesthetically pleasing	Style and visuals follow branding guidelines.
	Format (layout, spacing, grouping, and alignment)	The content is always conveyed in a consistent and compatible format to the user. The system adopts a visual hierarchy, mapping the relationships between controls and actions apparent to the user.
	Legibility and readability	Font sizes, font type, button sizes, and icons sizes enhance legibility and readability.
	Meaningful schematics, pictures, icons, and color	System provides schematics, pictures, and icons that are easy to understand to the user and adhere to effective usage of color.

Source: After Dharwada, P., and Tharanathan, A. 2011. Usability scorecard: A computational method for expert evaluation. In *Proceedings of the 2011 Industrial Engineering Research Conference*, ed. T. Doolen and E. Van Aken.

TABLE 14.2

Sample Finding from Heuristic Evaluation Scorecard

Finding	Prototype Iteration Number	Screen	Usability Area	Heuristic Violated	Finding Owner
Unconventional icon used to indicate unacknowledged alarm	3	Alarms	Visual design	System provides schematics, pictures, and icons that are easy to understand and adhere to effective usage of color.	TP

Including the risk ratings into a more quantitative risk score increases the precision of the heuristic evaluation findings. The most significant usability problems will contribute more heavily to the risk score for each heuristic and area than the less significant problems. The complete computational formula used by Dharwada and Tharanathan (2011) to compute risk scores is presented in Appendix 14.1.

Examples of this computation with real findings are shown in Tables 14.3 and 14.4. Table 14.3 shows that four usability problems were found in the area of Navigation. The problems violated the heuristic of "clear signs, cues, and aids for navigation and orientation." For each of the four problems, the evaluator assigned rating scores (9, 3, or 1) to each of three risk parameters (severity, frequency, detectability). For example, the problem noted as "confusion between BACK and CANCEL" was assigned ratings of 1 on severity, 1 on frequency, and 1 on detectability, or as shown in the last three columns of the table, "three 1s." Across the four problems, there were four ratings of 9, one rating of 3, and seven ratings of 1.

These totals are carried over into Table 14.4 which shows how they are transformed into a total risk score for the heuristic. As Table 14.4 shows, the first step is to compute a Cumulative Risk Score that is the sum of all the risk ratings, 12 ratings in this example. It is also an opportunity to apply a weighting factor to accentuate the particularly bad usability problems. The weights used as multipliers in the example are 0.6 for ratings of 9, 0.3 for ratings of 3, and 0.1 for ratings of 1. The Cumulative Risk Rating in the example is 23.2.

Next a Proportional Risk Rating is computed as Cumulative Risk Rating/ Maximum Possible Risk Rating. The maximum risk rating for a single problem would occur if a rating of 9 were assigned to all three risk parameters, giving a maximum possible risk rating of 27 for a single problem. In our example, there are four problems. If each was assigned the maximum possible risk rating of 27, it would total 108. Thus, 23.2/108 gives us the Proportional Risk Rating in this example, 0.2148. This is normalized by subtracting from 1, giving a normalized value of 0.785. Note that with this normalization, the higher the score, the lower is the usability risk.

The final step in the computation is to add a correction factor to the Proportional Risk Rating to account for the number of problems found under this single heuristic.

TABLE 14.3
Example of Risk Ratings and Computation of a Usability Risk Score for One Heuristic

Usability Area	Heuristic	Findings/Problems	Problem Risk Ratings			Total Number of Problems	Total Number of Risk Ratings		
			Severity	Frequency	Detectability		9s	3s	1s
Navigation	Clear signs, cues, and aids for navigation and orientation	Cannot access all setup options from View/Edit setup	9	1	9		2	0	1
		Users cannot go back directly to home when they are inside any screen	3	9	1		1	1	1
		Confusion between BACK and CANCEL	1	1	1		0	0	3
		Breadcrumbs are not clickable	1	9	1		1	0	2
						4	**4**	**1**	**7**

TABLE 14.4

Example of Risk Ratings and Computation of a Usability Risk Score for One Heuristic

Total Number of Problems (n)	Total Number of Risk Ratings			Cumulative Risk Score (Weighted)	Maximum Possible Risk Score	Proportional Risk Rating for the Heuristic	Normalized Risk Rating for the Heuristic	Corrected for Number of Issues per Heuristic (Four Problems, so Correction Factor is $n/2.75 = 1.45$)	Percent Score
	9s	3s	1s						
4	4	1	7	$4 * 9 * 0.6 + 1 * 3 * 0.3 + 7 * 1 * 0.1 = 23.2$	4 problems $*27 = 108$	$23.2/108 = 0.2148$	$1 - 0.2148 = 0.785$	$(0.785)^{1.45} = 0.7034$	70.34%

The correction is determined from the rules below and applied as an exponent to the normalized Proportional Risk Rating value:

$$Sh_i = (1 - Ph_i)^{m_i}$$

where i = heuristic i
$m = n_i/3$, if $n_i = 1$ or 2
$m = n_i/2.75$, if $n_i = 3$ or 4
$m = n_i/2.5$, if $n_i = 5$ or $n_i = 6$
$m = n_i/2.25$, if $n_i = 7$ or $n_i = 8$
$m = n_i/2$, if $n_i = 9$ or $n_i = 10$
$m = n_i/1.5$, if $n_i > 10$

In our example, the heuristic has four problems, so the correction factor is 4/2.75 = 1.45. Applied to the normalized Proportional Risk Rating of 0.785, we see that the score actually is slightly reduced to a value of 0.7034. The net effect of this correction is to produce a more negative ("riskier") rating as the number of problems increases and a more positive risk score as the number of problems decreases. Thus, if two heuristics have the same proportional score, but one has more problems than the other, the more problematic nature of that heuristic will be reflected in a more negative (lower) final rating.

14.2.2.2 Reporting

Once scores are computed using either of the above methods, numerous comparisons can be made to provide insight into design problem areas and iterative progress toward remediating them. Scores can be compared between usability areas and heuristics, and between iterations of the prototype so the development team can understand how the usability of the product is being improved over time.

14.3 CROSS-CULTURAL HEURISTIC EVALUATION

After seeing the scorecard above for conducting heuristic evaluation, it is clear that one could extend it by adding cross-cultural design guidelines as heuristics under each of the seven categories in which there might be cross-cultural design issues. In Table 14.5 we mapped the design guidelines presented in Section 2 of this book into eight areas of usability. We added an eighth usability area to the scorecard scheme to account for products that have ergonomic/anthropometric issues associated with them. This mapping allows us to use the cross-cultural design guidelines to evaluate the cross-cultural "goodness" of a design and actually give it a cross-cultural goodness score. We believe that it provides a practical way to use the guidelines in assessing progress on cross-cultural usability of the design throughout the development process. It can be used qualitatively to guide the evaluator in discovering cross-cultural design problems or more quantitatively, with risk ratings and scores assigned to each area and cross-cultural heuristic guideline. It is important to note that applying many of these guidelines as heuristics will require the use of a human factors expert with knowledge of the targeted local language and culture.

TABLE 14.5

Mapping of Cross-Cultural Usability Guidelines to Usability Areas and General Usability Heuristics

Usability Area	Usability Heuristics	Description	Cross-Cultural Usability Heuristics
Content	Relevant and precise content	The system provides content of information that is relevant to the user's context and presents in appropriate areas.	Translation and multiple language support of content: 4.2.5 Provide multiple language support 4.2.6 Make sure content matches the concepts of the selected language and values of the culture 4.2.4 Make sure words and concepts are translated to an appropriate context 4.2.2 Use technology jargon words carefully 4.2.7 Adapt to regional language preferences when designing speech interactions 4.2.1 Use Simplified English rendering language and number systems 4.2.3 Do not use abbreviations 4.2.12 Avoid the use of case as a distinguishing feature of characters 4.2.14 Use correct linguistic boundaries, ligatures, text wrappings and justifications, punctuation, diacritic marks, and symbols 4.2.15 Consider legibility factors when rendering text using Chinese characters
Functionality	Unambiguity/ appropriateness/ availability of required functionality	The system provides required functionality to the user. Specific functionality of the system is appropriate and useful and there is no ambiguity to the user.	4.2.16 Select an efficient text input method, particularly when Chinese characters must be entered 11.2.1 Ensure that the functions are sufficient and relevant to the user's culture, environment, and goals

Continued

TABLE 14.5 (Continued)
Mapping of Cross-Cultural Usability Guidelines to Usability Areas and General Usability Heuristics

Usability Area	Usability Heuristics	Description	Definitions — Cross-Cultural Usability Heuristics
Information architecture	Menu structures and hierarchy	Structuring of information into menus matches the user's mental model. Menu items, structure, and hierarchy are intuitive and easy to learn.	7.2.2 For menu design, provide orientation compatible with the language being presented 4.2.10 Use an appropriate method of sequence and order in lists 8.2.1 Information and functions should be organized in a manner consistent with the target user's natural way of grouping functions, concepts, and objects
	Organization/home screen layout	Home screen provides the user with clear purpose of the system and defines the organization of the application. User actions on the system are made clear and give direct access to key content/tasks (search box, etc.).	8.2.1 Information and functions should be organized in a manner consistent with the target user's natural way of grouping functions, concepts, and objects 8.2.5 Provide extra navigational aids for Japanese, Arabic, and Mediterranean users or users in high-context communication style
Navigation	Avoid deep navigation	Key content and functions are visible in the first screenshot and do not require scrolling or deeper levels of navigation.	
	Clear signs, cues, and aids for navigation and orientation	The system provides clearly visible/understandable elements that help users in orientation within the system navigation and context. The system provides navigation aids to efficiently navigate through long lists, trace back the path taken, or return to the primary context within a task.	8.2.2 Provide searching mechanisms. Ensure that the functions are sufficient and relevant to the user's culture, environment, and goals 8.2.3 Provide both search engine and Web directory to support users from different cultures with different search mechanism preferences 8.2.5 Provide extra navigational aids for Japanese, Arabic, and Mediterranean users or users in high-context communication style

Continued

System responsiveness	Highly responsive	System responds very quickly, user does not have to wait too long for the system to provide response.	
User control and freedom	Error prevention, recovery, and control	Even better than good error messages is designing for prevention of a problem from occurring in the first place. Either eliminate error-prone conditions or check for them and present users with a confirmation option before they commit to the action. Users often choose system functions by mistake and will need a clearly marked "emergency exit" to leave the unwanted state without having to go through an extended dialogue (Recovery).	
	Flexibility, control, and efficiency of use	The system provides accelerators—unseen by the novice user—may often speed up the interaction for the expert user such that the system can cater to both inexperienced and experienced users. Allow users to tailor frequent actions. Provides additional quick ways of supporting user workflow practices.	
User guidance and support	Consistency	The system follows standard user conventions across the application. Users should not have to wonder whether different words, situations, controls, or actions mean the same thing. The system behavior corresponds to user expectations.	
	Compatibility	Match between real-world and system status. Compatibility with metaphors chosen for presentation.	
	Informative feedback and status indicators	The system always keeps users informed about what is going on through appropriate feedback and when needed provides with course of action within reasonable time.	8.2.4 Provide possible outcomes and results of operations as much as possible for Asian users or users in high-uncertainty-avoidance cultures

TABLE 14.5 (Continued)
Mapping of Cross-Cultural Usability Guidelines to Usability Areas and General Usability Heuristics

		Definitions	
Usability Area	Usability Heuristics	Description	Cross-Cultural Usability Heuristics
	Recognition rather than recall	Minimize the user's memory load by making objects, actions, and options visible. The user should not have to remember information from one part of the dialogue to another. Instructions for use of the system should be visible or easily retrievable whenever appropriate.	
	Terminology: informative titles, labels, prompts, and messages	The system speaks the user's language with words, phrases, and concepts familiar to the user, rather than system-oriented terms. Follow real-world conventions, making information appear in a natural and logical order.	
	Tool tips, help, and documentation	The system provides tool tips where necessary. Even though it is better if the system can be used without documentation, it may be necessary to provide help and documentation. Any such information should be easy to search, focused on the user's task, list concrete steps to be carried out, and not be too large.	
	Workflow support	System work-flow is compatible to user mental model of the workflow and guides the user along the workflow.	
Visual design	Aesthetically pleasing	Style and visuals follow branding guidelines.	5.2.2 Use colors that are affectively satisfying in the target culture

Format (layout, spacing, grouping, and alignment)	The content is conveyed always in a consistent and compatible format to the user. The system adopts visual hierarchy, mapping the relationships between controls and actions apparent to the user.	4.2.8 Allow extra space for text 4.2.11 Avoid combining user interface (UI) objects into phrases 4.2.13 Text directionality
Legibility and readability	Font sizes, font type, button sizes, and icon sizes enhance legibility and readability.	4.2.15 Consider legibility factors when rendering text using Chinese characters 7.2.1 Provide natural layout orientation for information to be visually scanned
Meaningful schematics, pictures, icons, and color	System provides schematics, pictures, and icons that are easy to understand to the user and adheres to effective usage of color.	Icons: 6.2.1 Make sure icons are highly recognizable to the target users 6.2.2 When designing icons, provide a combination of text and picture 6.2.5 Avoid using graphics with culture specific metaphors and associations 6.2.6 Make use of appropriate symbols, images, graphics, and colors that are highly recognized in the target culture. 6.2.7 Insure that graphics reflect the dominant social values of the target locale 4.2.9 Do not embed text within icons Graphics: 6.2.3 Make sure the textual components of graphics are compatible with the language(s) of the target users 6.2.4 If possible, avoid using graphics with culture-specific meanings or associations Color: 9.2.2.1 Color associations with safety conditions 9.2.2.2 Color and affect
Anthropometric design		9.2.3.1 Select suitable area database for cross-cultural design 9.2.3.2 Pay attention to area difference of anthropometric data within one country

An example of using the cross-cultural design guidelines in a heuristic evaluation of a website is provided in the case study below.

14.4   CASE STUDY: CROSS-CULTURAL HEURISTIC EVALUATION OF AN E-COMMERCE WEB SITE

14.4.1   BACKGROUND

E-commerce websites have become very popular in China. Many Chinese e-commerce websites are simply subsites of international e-commerce websites. International e-commerce websites establish or purchase local e-commerce websites and may attempt to bring their successful experiences, including a user interface design. Thus, many such e-commerce websites use more or less the same user interface design across cultures, with the major accommodation being the translation of text to the local language.

There are also many successful local Chinese e-commerce websites. In the beginning, they learned and copied successful business models and designs from international e-commerce websites, particularly those from the United States. Now, most local Chinese e-commerce websites design the user interface from local experiences and wisdom by paying attention to local cultural characteristics and local users' requirements.

14.4.2   OBJECTIVE

This case study was aimed at evaluating the cross-cultural usability of a typical international e-commerce website deployed in China. Also, this case study served to confirm the usefulness of the cross-cultural heuristic guidelines summarized in this book.

14.4.3   METHODOLOGY

A leading international e-commerce website, based in the U.S., and also invested in and operated in China, was chosen for study. For proprietary reasons, throughout this case study, the company will be referred to simply as "Shoppingbiz.cn" for its China operation, or "Shoppingbiz.com" for its U.S. operation. The web design of Shoppingbiz.cn is consistent with that used in Shoppingbiz.com and only differs in language.

The method of cross-cultural heuristic evaluation was conducted as summarized in previous chapters of this book. The evaluation process contained the following four steps.

14.4.3.1   Task Design

First, three scenarios of use were defined. The specific user tasks were identified that would be performed frequently by common users, from registration to final payment.

14.4.3.2   Expert Evaluation

Human factors experts experienced the three use scenarios and finished the three tasks individually. Usability problems of the evaluated e-commerce website were

TABLE 14.6
Principles of Rating Three Risk Criteria

Risk Criteria	Risk Level	Principle
Severity	Low risk (1)	This problem exists but merely influences users to finish their tasks.
	Medium risk (3)	Sometimes users will be puzzled by this problem, and they need to make some effort to finish their tasks.
	High risk (9)	This problem causes great difficulty to users, and they need to make great effort to understand how to finish their tasks. Even they will give up trying and leave.
Frequency of occurrence	Low risk (1)	This problem merely occurs.
	Medium risk (3)	This problem occurs in some pages or in some situations.
	High risk (9)	This problem occurs in almost every page or every task.
Detectability	Low risk (1)	Most times, users are not aware of this problem.
	Medium risk (3)	In some situations, users realize that they encounter a usability problem that bothers their operation.
	High risk (9)	User is always aware of this problem.

identified during and after carrying out these tasks. When they were doing the tasks, the human factors experts thought aloud to express their findings. What's more, a voice recorder and screen recording software were used to record this process. In addition, a research assistant recorded key notes on paper.

14.4.3.3 Identification and Classification

After expert evaluation, we made a list of all usability problems and then checked and identified the findings according to audio and video records and the assistant's notes. Finally, the findings were classified in accordance with Cross-cultural Usability Heuristics Guidelines.

14.4.3.4 Score

In the last step, we summarized the usability problems and scored the findings, then calculated the final heuristic score for each guideline. We rated each usability problem with three risk criteria, which include Severity, Frequency of Occurrence, and Detectability. Then we calculated the total score for each guideline with the heuristic evaluation scorecard designed by Dharwada and Tharanathan (2011). Table 14.6 presents the principles of rating the three risk criteria for each usability problem.

14.4.4 RESULTS AND ANALYSIS

In the heuristic evaluation, 38 cross-cultural usability problems were found. They were spread across 10 cross-cultural heuristic guidelines. The final heuristic scores of these guidelines were as follows.

From Table 14.7, we can see that four guidelines (8.2.1, 7.2.1, 11.2.1, and 4.2.15) received very low heuristic scores. There are two reasons. First, usability problems

TABLE 14.7

Result of Cross-Cultural Heuristic Evaluation for Shoppingbiz.cn

Cross-Cultural Usability Heuristics	Number of Problems	Total Multiplication with Coefficients	Percent (%) Heuristic Score
4.2.4	3	13.5	82.0
4.2.16	1	6.3	91.5
11.2.1	6	58.5	**34.1**
8.2.1	7	78.3	**18.9**
8.2.4	3	28.8	61.9
6.2.1	1	2.7	96.5
4.2.8	1	2.7	96.5
7.2.1	6	75.6	**22.1**
4.2.15	4	45.9	**44.7**
6.2.1	1	8.1	88.8

that violated these guidelines have high risk levels (i.e. severity, frequency of occurrence, and detectability), and second, the number of usability problems is high.

14.4.4.1 Analysis of Guideline 8.2.1

Information and functions should be organized in a manner consistent with the target user's natural way of grouping functions, concepts, and objects.

Seven problems were found that violated guideline 8.2.1. The classification and categories of products in Shoppingbiz.cn follow Shoppingbiz.com. However, in some places, their information architecture fails to meet Chinese users' requirements.

One typical problem we found is that when we point at one item in the product category of Shoppingbiz.cn (on the left side of page), there are just a few items appearing in the second-level category. In contrast, in the other most popular local online shopping websites in China, users can find many more items in the second-level category, and they can easily choose the specific kinds of products they prefer. Figure 14.1, second-level categories of products, illustrates this.

We can use the principle introduced above to rate the three risk criteria for this problem. This problem forces users into more steps to find what they want. Eventually, most users find their preferred products, but it takes more steps. Thus, we assigned the problem to a medium risk level (three points) on severity. Almost always, the user will use the product categories to search for the product they want. It is almost unavoidable. Therefore, we assigned the problem a high risk level (nine points) on frequency of occurrence. At the beginning, users may have difficulty finding the preferred product. But after using this website several times, they become familiar with the information architecture and can find the product easily. Thus the problem has medium risk level (three points) on Detectability.

When the users finally chose a specific category of product and entered the subpage for this category, they often wanted to set some conditions to filter the product

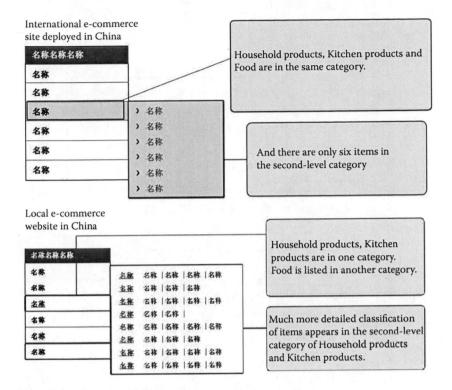

FIGURE 14.1 Second-level category of products.

list. For some products in Shoppingbiz.cn, there was no help for users to filter on the category and find their preferred products. For instance, in the page of jeans in Shoppingbiz.cn, the first three filter conditions refer to color, size, and brand. But when the Chinese user browses jeans online, what they care about most are brand, price, style, and material. Shoppingbiz.cn also provides the filter conditions about price, but users have to scroll the screen to find it, making it difficult to find. Shoppingbiz.cn does not allow users to filter products by style and material for jeans. Figure 14.2, screen conditions for selecting jeans, illustrates this point.

14.4.4.2 Analysis of Guideline 7.2.1

Provide natural layout orientation for information to be visually scanned.

Six usability problems violated guideline 7.2.1. The layout of information in Shoppingbiz.cn is the same as Shoppingbiz.com, without considering the differences between English and Chinese (e.g., length of a word, complexity of character, text directionality, and the smallest allowed font size). The literal translation of Shoppingbiz.com's layout violates the reading habits of Chinese users, making it difficult for Chinese users to extract important information from a group of text.

One typical usability problem related to information layout in Shoppingbiz.cn is the information layout of customer comments. In Shoppingbiz.cn, it simply differentiates

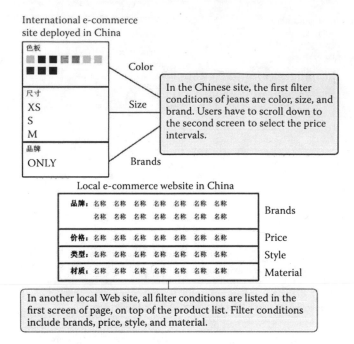

International e-commerce
site deployed in China

In the Chinese site, the first filter conditions of jeans are color, size, and brand. Users have to scroll down to the second screen to select the price intervals.

Local e-commerce website in China

In another local Web site, all filter conditions are listed in the first screen of page, on top of the product list. Filter conditions include brands, price, style, and material.

FIGURE 14.2   Filter conditions of jeans (**see color insert**).

the information by changing the size, color, and font of the text. When users browse these texts, it is very hard for them to extract useful content. Compared with English, Chinese uses many fewer characters and less space to present the same amount of information, Thus, the information layout design in Chinese can be more hierarchical. Figure 14.3, information layout of customer comments, shows this.

Another example is the format of product information on the right side of the product image. Here, so many different fonts, character sizes, text colors, and alignments are applied that people find it hard to pick up the most useful information on their first attempt. Figure 14.4 illustrates the product information with various alignments, fonts, character sizes, and colors.

14.4.4.3   Analysis of Guideline 11.2.1

Ensure that the functions are sufficient and relevant to the user's culture, environment, and goals.

Six usability problems violated guideline 11.2.1. Shoppingbiz.cn is not aware of some functions and workflows that differ between the United States and China. During the evaluation, we found that some functions in Shoppingbiz.cn are not well fitted to Chinese users.

One example, shown in Figure 14.5, is that when users try to place an order, they must enter the password again to verify their identity. This function may be necessary in the United States, as the product will be paid via credit card directly

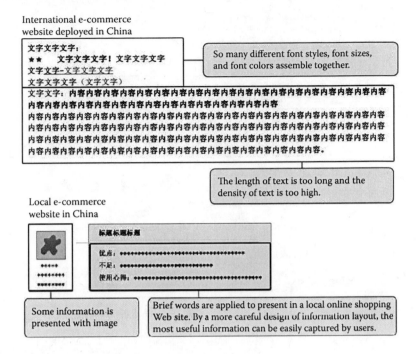

FIGURE 14.3 Information layout of customer comments.

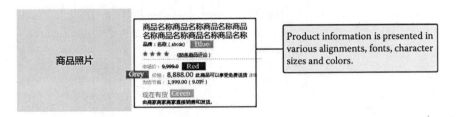

FIGURE 14.4 Product information with various alignments, fonts, character sizes, and colors (**see color insert**).

without any other operations. However, in China the process of online payment is not so convenient. There are still many steps required for users to verify their identities. Figure 14.6 illustrates this problem of redundant password entry and why the requirement should be removed for Chinese users.

14.4.4.4 Analysis of Guideline 4.2.15

Legibility factors are very important when rendering text using Chinese characters.

Four usability problems violated guideline 4.2.15. Because the complexity of Chinese characters is much higher than English characters, the smallest allowed sizes of Chinese characters for good legibility is different. We found that in most cases, the reading speed of images was faster than the text. More images could be

FIGURE 14.5   Users needs to login again when placing an order.

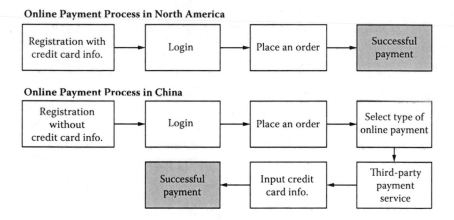

FIGURE 14.6   Online payment processes in the United States and China.

used in the Shoppingbiz.cn website to present the products, rather than only text. Figure 14.7 shows the low legibility of information.

14.5   DISCUSSION

The heuristic evaluation results found that the design of the Chinese version of the Shoppingbiz website did not satisfy the requirements of local users for online shopping, especially in the following four aspects:

1. Information and functions were not organized in a manner consistent with the target user's natural way of grouping functions, concepts, and objects.
2. Information layout was not well customized to the regional user's natural visual scan habits.

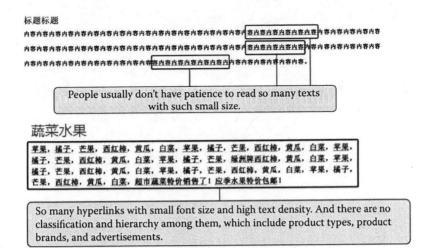

FIGURE 14.7   Low legibility of information.

3. Certain functions, relevant in the United States were not relevant to the Chinese user's culture, environment, and goals.
4. Legibility and readability of information presented on the website sometimes suffered from rendering the Chinese characters with insufficient font size.

When browsing and buying things on Shoppingbiz.cn, new users may encounter usability problems that reduce their user experience and may even cause them to give up trying to select a product and make a purchase. In this evaluation, only three scenarios were tested by two experts. So only a subset of the cross-cultural usability problems was identified. More extensive testing of other scenarios of use and tasks would be required to cover all the guideline areas.

APPENDIX 14.1: HEURISTIC SCORECARD COMPUTATIONAL FORMULA (DHARWADA AND THARANATHAN, 2011)

n_i = Total number of violations or issues per heuristic

$$n_i = \sum_{i=heuristic_i} issue_i$$

x_i = total number of risk ratings that equal a value of nine across the three risk dimensions (risk severity, occurrence, and detectability) on all the issues identified for heuristic

y_i = total number of risk ratings that equal a value of three across the three risk dimensions (risk severity, occurrence, and detectability) on all the issues identified for heuristic$i$

z_i = total number of risk ratings that equal a value of one across the three risk dimensions (risk severity, occurrence, and detectability) on all the issues identified for heuristic$i$

$$x_i = \sum_{i=heuristici,\,severity=9} issue_{i,severity} + \sum_{i=heuristici,\,occurence=9} issue_{i,occurence} + \sum_{i=heuristici,\,detectability=9} issue_{i,detectability}$$

$$y_i = \sum_{i=heuristici,\,severity=3} issue_{i,severity} + \sum_{i=heuristici,\,occurence=3} issue_{i,occurence} + \sum_{i=heuristici,\,detectability=3} issue_{i,detectability}$$

$$z_i = \sum_{i=heuristici,\,severity=1} issue_{i,severity} + \sum_{i=heuristici,\,occurence=1} issue_{i,occurence} + \sum_{i=heuristici,\,detectability=1} issue_{i,detectability}$$

Proportion for heuristic

$$Ph_i = \frac{(9 \times 0.6)x_i + (3 \times 0.3)y_i + (1 \times 0.1)z_i}{27n_i}$$

where $i = heuristici$.

Score per heuristic

$$Sh_i = (1 - Ph_i)^{m_i}$$

where

$i = heuristici$

$m_i = n_i / 3$, if $n_i = 1$ or 2

$m_i = n_i / 2.75$, if $n_i = 3$ or 4

$m_i = n_i / 2.5$, if $n_i = 5$ or $n_i = 6$

$m_i = n_i / 2.25$, if $n_i = 7$ or $n_i = 8$

$m_i = n_i / 2$, if $n_i = 9$ or $n_i = 10$

$m_i = n_i / 1.5$ if $n_i > 10$

Proportion for area

$$Pa = \frac{(9 \times 0.6)\sum_{i=1}^{l} x_i + (3 \times 0.3)\sum_{i=1}^{l} y_i + (1 \times 0.1)\sum_{i=1}^{l} z_i}{\sum_{i=1}^{l} 27 n_i}$$

where

$$l = \sum_{heurisic \in usability\ area} heuristic$$

Score per area

$$Sa = (1 - Pa)^m$$

where

$$m = \frac{\sum_{i=1}^{l} n_i}{3}, \quad \text{if } \sum_{i=1}^{l} n_i = 1 \text{ or } 2$$

$$m = \frac{\sum_{i=1}^{l} n_i}{2.75}, \quad \text{if } \sum_{i=1}^{l} n_i = 3 \text{ or } 4$$

$$m = \frac{\sum_{i=1}^{l} n_i}{2.5}, \quad \text{if } \sum_{i=1}^{l} n_i = 5 \text{ or } \sum_{i=1}^{l} n_i = 6$$

$$m = \frac{\sum_{i=1}^{l} n_i}{2.25}, \quad \text{if } \sum_{i=1}^{l} n_i = 7 \text{ or } \sum_{i=1}^{l} n_i = 8$$

$$m = \frac{\sum_{i=1}^{l} n_i}{2}, \quad \text{if } \sum_{i=1}^{l} n_i = 9 \text{ or } \sum_{i=1}^{l} n_i = 10$$

$$m = \frac{\sum_{i=1}^{l} n_i}{1.5}, \quad \text{if } \sum_{i=1}^{l} n_i > 10$$

$$l = \sum_{heurisic \in usability\ area} heuristic$$

Score per area percentage score

$$PSa = Sa \times 100$$

Defect rate/defect density

$$d = \frac{\sum screen}{\sum issue}$$

Overall score

$$Overall\ Score = \begin{cases} \dfrac{\sum\limits_{i=1}^{k} PSa_i}{k}, & d \geq 1 \\[2em] \dfrac{1.75 \sum\limits_{i=1}^{k} PSa_i}{k \cdot d}, & d < 1 \end{cases}$$

where

$$k = \sum usability\ area$$

REFERENCES

Burns, C.M., and Hajdukiewicz, J.R. 2004. *Ecological Interface Design*. Boca Raton, FL: CRC Press.

Dharwada, P., and Tharanathan, A. 2011. Usability scorecard: A computational method for expert evaluation. In *Proceedings of the 2011 Industrial Engineering Research Conference*, ed. T. Doolen and E. Van Aken.

Nielsen, J. 1993. *Usability Engineering*. San Francisco, CA: Morgan Kaufmann.

Nielsen, J. 1994. Heuristic evaluation. In *Usability Inspection Methods*, ed. J. Nielsen and R.L. Mack, 25–62. New York: Wiley.

Appendix 1: Guidelines Ratings

Chapter	Original Guidelines	Category	Source Category	Source 1	Source 2	Source 3
Chapter 4: Language	Use Simplified English	Guideline	Design guidelines	AECMA (Association Europeenee des Constructeurs de Material Aerospatial)	Mills and Caldwell (1997)	IBM (2012a)
	Use technology jargon words carefully	Guideline	Design guidelines and experiments	Zhang, Zhao, and Zhang (2000)	IBM (2012a)	Oracle (2010)
	Do not use abbreviations	Guideline	Design guidelines and experiments	Röse et al. (2001)	IBM (2012a)	
	Make sure words are translated to an appropriate context	Guideline	Design guidelines	IBM (2012b)		
	Provide multiple language support	Guideline	Design guidelines	Luna et al. (2002)	Peracchio and Meyers-Levy (1997)	IBM (2012b); MicroSoft (2012); W3C (2012)
	Make sure content matches the concepts and values of the selected language	Guideline	Design guidelines	De Groot (1991)	Luna et al. (2002)	W3C (2012)
	Adapt to regional preferences when designing speech interactions	Research	Experiments	Chan and Khalid (2000)		
	Allow extra space for text	Guideline	Design guidelines	IBM (2012b)	MicroSoft (2012)	Oracle (2010)
	Do not embed text in icons	Guideline	Design guidelines	IBM (2012b)	MicroSoft (2012)	Oracle (2010)
	Use an appropriate method of sequence and order in lists	Guideline	Design guidelines	IBM (2012b)	MicroSoft (2012)	Oracle (2010)

Chapter	Guideline					
	Avoid combining UI objects into phrases	Guideline	Design guidelines	MicroSoft (2012)	Oracle (2010)	
	Avoid the use of case as a distinguishing feature of characters	Guideline	Design guidelines	MicroSoft (2012)		
	Text directionality	Guideline	Design guidelines and experiments	Goonetilleke, et al. (2002)	MicroSoft (2012)	Oracle (2010); W3C (2012)
	Use correct linguistic boundaries, ligatures, text wrappings and justifications, punctuation, diacritic marks, and symbols	Guidelines	Design guidelines	IBM (2012b)	MicroSoft (2012)	
	Consider legibility factors when rendering text using Chinese characters	Research findings	Experiments	Jin, et al.(1988)	Cai, et al. (2001)	Shieh, et al. (1997)
	Select an efficient text input method when Chinese characters must be entered	Research findings	Experiments	Niu, et al. (2010)		
Chapter 5: Color coding and affect	Color Associations With Safety Conditions	Guidelines	Design guidelines and experiments	ASM (2008)	Courtney (1986); Luximon, et al. (1998)	Liang, et al. (2000); Liang, et al (2004); Kaiser (2002)
	Color and Affect	Research findings	Experiments	Spartan (1999); Osgood, et al. (1975)		

Continued

Chapter	Original Guidelines	Category	Source Category	Source 1	Source 2	Source 3
Chapter 6: Icons and Images	Make sure icons are highly recognizable to the target users	Guideline	Design guidelines and experiments	IBM (2012b), MicroSoft (2012); Oracle (2010)	Shen, et al. (2007); Choong, et al. (2010); Pappachan et al. (2008); Kim et al. (2005)	Piamonte, et al. (1997, 1999); Plocher et al. (1999)
	When designing icons, provide a combination of text and picture	Research findings	Experiments	Choong and Salvendy (1998)	Kurniawan, et al. (2001)	
	Make sure the textual components of graphics are compatible with the language(s) of the target users	Guideline	Design guidelines	IBM (2012b)		
	Design graphics to support natural reading and scanning direction	Theoretical	Design Implications (un-validated)	Horton (1994)		
	Avoid using graphics with culture-specific metaphors and associations	Guideline	Design guidelines	Fernandes (1995)	JBM (2012)	MicroSoft (2012); Oracle (2010)

	Guideline	Type	Category	Reference	Reference	Reference
	Make use of appropriate symbols, images, graphics, and colors that are highly recognized in the target culture to excite and please the user.	Guideline	Design guidelines	Luna et al. (2002)	Minocha et al. (2002)	Fernandes (1995)
	Ensure that graphics reflect, or at least do not contradict, the dominant social values of the target locale for social distance, point of view, degree of involvement, and power.	Theoretical	Design Implications (un-validated)	Gould (2004, 2004)		
Chapter 7: Presentation, Format and Layout	Provide natural layout orientation for information to be scanned.	Research findings	Experiments	Lau, Goonetilleke,, et al. (2001)	Goonetilleke, et al. (2002)	
	For menu design, provide orientation compatible with the language being presented.	Research findings	Experiments	Shih, et al. (1997, 1998)	Dong, et al. (1999)	
	Text Direction, Labeling and Scrolling	Guideline	Design guidelines	IBM (2012b)	MicroSoft (2012)	

Continued

Chapter	Original Guidelines	Category	Source Category	Source 1	Source 2	Source 3
Chapter 8: Information Organization and Representation/ Information Structure (navigation and hyperlinks)	Information should be organized in association with target user's cultural traits	Research findings	Experiments	Choong (1996, 1999)	Luna, et al. (2002)	Rau, et al. (2004)
	Provide searching mechanisms. Ensure that the functions are sufficient and relevant to the user's culture, environment, and goals	Guidelines	Design guidelines	Morkes and Nielsen (1997)	Nielsen (1997)	Zhao (2002)
	Provide both search engine and web directory to support different needs of users.	Research findings	Experiments	Fang and Rau (2003)		
	Provide possible outcomes and results of operations as much as possible for Asian users or users in high uncertainty avoidance cultures.	Theoretical	Design Implications (un-validated)	Marcus et al., (2000)	Plocher, et al. (1999)	

			Cyr and Trevor-Smith, (2004)	Rau and Liang (2003a, 2003b)	Plocher et al. (2001)
Provide extra navigational aids for Japanese, Arabic, and Mediterranean users or users in high-context communication style.	Research findings	Experiments			
Chapter 9 Physical Ergonomics and Anthropometry					
Select a suitable anthropometry database for cross-cultural design	Guideline	Design guidelines	1988 Anthropometric Survey of US Army Personnel (ANSUR); Air Force surveys; National Health and Nutrition Examination Survey (NHANES); Civilian American and European Surface Anthropometry Resource (CAESAR); Size UK ERGODATA; Size China; Size Japan; WEAR; DINED; AnthroKids; DINBelg 2005		
Pay attention to area differences in anthropometric data within one country	Guideline	Design guidelines	GB 10000-88 (China)		

Appendix 2: Cross-Cultural Psychology and HCI Resources

1. http//www.culturalstudies.net/
2. http://www.uiowa.edu/~commstud/resources/culturalStudies.html
3. http://www.pcaaca.org/
4. http://theory.eserver.org/
5. http://www.popmatters.com/popcultures/
6. http://www.theory.org.uk/
7. http://www.iaccp.org/
8. http://www.sccr.org/
9. http://www.itopwebsite.com/InternationalPsychology/HOME.html
10. http://www.wwu.edu/culture/contents_complete.htm
11. Ben Shneiderman and Catherine Plaisant, *Designing the User Interface: Strategies for Effective Human-Computer Interaction*, 5th edition, Pearson Addison-Wesley, Reading, MA, 2009.
12. Donald A. Norman, *The Design of Everyday Things*, Basic Books, New York, 2002.
13. Jeff Johnson, *Designing with the Mind in Mind: Simple Guide to Understanding User Interface Design Rules*. Morgan Kaufmann, 1st Edition (June 3, 2010).
14. Alan Cooper, *About Face 3: The Essentials of Interaction Design*, 3rd edition, Wiley, New York, 2007.
15. Alan Cooper, *The Inmates Are Running the Asylum: Why High Tech Products Drive Us Crazy and How to Restore the Sanity*. Sams—Pearson Education, 1st Edition (March 5, 2004).
16. Bill Moggridge, *Designing Interactions*, MIT Press, Cambridge, MA, 2008.
17. Jenny Preece, Yvonne Rogers, and Helen Sharp, *Interaction Design: Beyond Human-Computer Interaction*, 3rd edition, Wiley, New York, 2011.
18. Bill Buxton, *Sketching User Experiences: Getting the Design Right and the Right Design (Interactive Technologies)*, Elsevier, New York, 2007.
19. Terry Winograd (Ed.), *Bringing Design to Software*, Addison-Wesley, Reading, MA, 1996.
20. Apple Computer, The Apple Software Design Guidelines, updated regularly.
21. Psychology and Cognitive Sciences
22. *American Journal of Psychology*
23. *Annual Review of Psychology*
24. *Applied Cognitive Psychology*
25. *Behavioral and Brain Sciences*
26. *Cognition*

27. *Cognitive Psychology*
28. *Cognitive Science*
29. *Cyber Psychology and Behavior*
30. *Journal of Applied Behavioral Science*
31. *Journal of Applied Psychology*
32. *Journal of Cognitive Psychology*
33. *Methods of Psychological Research* (online)
34. *Psychological Bulletin*
35. *Small Group Research*
36. *ACM Interactions*
37. *Advances in Human-Computer Interaction*
38. *Behavior Research Methods, Instruments, and Computers*
39. *Behavior and Information Technology*
40. *Computers in Human Behavior*
41. *Decision Support Systems*
42. *HCI Journal of Information Development*
43. *Human Technology* (online)
44. *International Journal of Human-Computer Interaction*
45. *Journal of Cognitive Engineering and Decision Making*
46. *Journal of Computer-Mediated Communication*
47. *Minds and Machines*
48. *Multimedia Systems*
49. *The International Journal of Virtual Reality*
50. *User Modeling and User-Adapted Interaction*
51. *Applied Ergonomics*
52. *Ergonomics*
53. *Human Factors*
54. *International Journal of Industrial Ergonomics*
55. *ACM Communications*
56. *ACM Transactions on Software Engineering and Methodology*
57. *Design Studies*

Index

A

Abbreviations, 34, 35
ADULTDATA, 133
AECMA (Association Europeenee des Constructeurs de Material Aerospatial) Simplified English, 31, 33. *See also* Simplified English
Alphabet, phonetic, 20
Altruism, 10
Ambiguity, 35
American culture/users. *See also* English, American
   case study, cultural affects of knowledge representations for user interface for online shopping; *see* Case study, cultural affects of knowledge representations for user interface for online shopping
   cognitive style of, 121
   icon recognition, 74, 83–84, 85
   information structure preferred by, 127
Analysis of variance (ANOVA), 105
Anthropometry, 3, 10
   case study, helmet shell design; *see* Case study, anthropometric measurement, Chinese medial acupuncture; Case study, helmet shell design
   databases, data collection, 133, 136
   databases, importance of selection, 133, 136; *see also specific databases*
   defining, 131
   movement, human, 131–132
   overview, 131
   product design, influence on, 132
   reach zone, 132
   user population, importance of selection, 136–137
Apple products, 181
Arabic language, 19, 101
Aristotle, 13
Automated voices, 36, 37

B

Bias, cultural. *See* Cultural bias in moderator-user interactions
Bilingual mobile phones
   evaluation, 51–53
   heuristic criteria, 49, 50
   objectives, 49
   overview, 47, 49
   results, 53–57
   scripts, differences in, 50–51
Biometric technologies, case study. *See* Case study, biometric technologies

C

Cangjie, 45
Card Sort Analyzer, 113
CardZort, 113
Case study, anthropometric measurement, Chinese medial acupuncture
   bone length measurement, 139–140
   method, 142
   objectives, 142
   overview, 139–142
   results, 142
Case study, biometric technologies
   method, 87
   objective, 87
   overview, 85, 86–87
   procedure, 88
   recognition tests, 88, 89–90, 91, 92
Case study, helmet shell design
   conclusions, 138–139
   head models, 3D, 138
   methods, 138
   objectives, 137
   overview, 137
   results, 138
   user population, 138
Case study, home health-case in rural India
   data collection, 170
   findings, 174–175
   map making exercise, 170–171, 173, 174
   overview, 169–170
   participation, 170
Case study, photointerviews to understand independent living needs of elderly Chinese
   checklist, structured, 163, 165
   data collection, 163
   interview questions, structured, 165
   interviewers, 163
   object inventory, 166–167
   overview, 162
   participants, 162